POLLUTION PREVENTION

POLLUTION PREVENTION

Problems and Solutions

Edited by

Louis Theodore
Manhattan College, Riverdale, New York

R. Ryan Dupont
Utah State University, Logan

and

Joseph Reynolds
Manhattan College, Riverdale, New York

Gordon and Breach Science Publishers

USA   Switzerland   Australia   Belgium   France   Germany   Great Britain
India   Japan   Malaysia   Netherlands   Russia   Singapore

Gordon and Breach Science Publishers S.A.
World Trade Center
Case postale 531
1000 Lausanne 30 Grey
Switzerland

The development of this problem workbook was made possible by an Undergraduate Faculty Enhancement grant from the National Science Foundation (Grant # USE 9154224).

British Library Cataloguing in Publication Data

Pollution Prevention: Problems and
Solutions
 I. Theodore, Louis
 628.5

  ISBN 2-88449-142-2 (hardcover)
        2-88449-129-5 (softcover)

CONTENTS

PREFACE

The United States currently spends over $100 billion a year, or approximately 2% of its gross national product on pollution control. The United States Environmental Protection Agency (U.S. EPA) estimates that by the year 2000, the cost of such protection could rise to as high as $185 billion a year (in 1990 dollars), or 2.8% of the GNP. These findings are contained in *Environmental Investments: The Cost of a Clean Environment*, a report former U.S. EPA administrator William K. Reilly recently sent to Congress. More than two years in the writing, the "cost-of-clean" report was mandated by the Clean Water Act, but it is likely to strongly influence the direction that environmental protection takes in the coming decades.

Although the engineering and science profession has recently expanded its responsibilities to society to include the management of wastes, with particular emphasis on hazardous and toxic wastes, it has now become clear that the principal waste management option of the future will be pollution prevention (P^2) and/or waste minimization. Increasing numbers of engineers, scientists, administrators, and field personnel are being confronted with complex problems best solved through implementation of P^2 principles. At present, however, few practicing professionals have a working understanding of this approach, educators are not presenting course material in this field, and students are not being provided with the necessary training to implement P^2 principles in their future work environment. Since the problem of waste management/prevention via waste minimization is a relatively new concern, the engineers and applied scientists of today and tomorrow must develop a proficiency and an improved understanding of this waste management option in order to cope with the challenges of the new "green" U.S. and world economies.

Recognizing the need to support undergraduate educators in the development of P^2 educational materials, the National Science Foundation funded a College Faculty Workshop on Pollution Prevention that was conducted at Manhattan College in June 1992. NSF College Faculty Workshops are designed to "involve college faculty members in preparing course materials and . . . in testing the effectiveness of these science curricular innovations for implementation." To qualify, topics must be of sufficiently broad applicability and impact to warrant national implementation for the enhancement of undergraduate science curricula. In awarding College Faculty Enhancement grants, the NSF gives priority to the development of more efficient and effective educational procedures in newly emerging, interdisciplinary, and problem-relevant subject areas. The principal objectives of the Manhattan College seminar on P^2 were: (1) to provide a meaningful course which would generate new ideas and innovative educative approaches in the emerging field of Pollution Prevention, and (2) to develop an applications-oriented problem workbook which would support undergraduate faculty involved in the production of course materials in P^2.

An interdisciplinary group of engineering and science faculty from across the United States involved in undergraduate environmental engineering or applied science courses were selected to attend the two-week seminar. These faculty are listed in the Contributors section. Each faculty member was required to generate six meaningful, applications-oriented problems. The 1992-1993 academic year afforded workshop participants an opportunity to classroom test these problems. A three-day follow-up session held in June 1993 at Colorado State University was utilized to revise, update, and edit the workbook. *Pollution Prevention: Problems and Solutions* is the product of this year-long effort.

Chapter 1 of this book provides an overview of pollution prevention and life cycle cost analysis concepts. Chapter 2 contains more than 100 problems related to a variety of topics of relevance to the pollution prevention field. These problems are organized into the categories of: Basic Concepts, Pollution Prevention Principles, Regulations, Source Reduction, Recycling, Treatment / Disposal, Chemical Plant / Domestic Applications, Case Studies, Ethics, and Term Papers and Projects.

Detailed solutions to each problem are provided in Chapter 3. Solutions are identified with a number and name corresponding to the original problem given in Chapter 2; for example, the solution to Problem "5.2 Recycling—2" is found in Chapter 3 under the title "5.2 Recycling Solution—2."

Acknowledgments are due to Joseph Reynolds of Manhattan College and to Sonia Kreidenweis of Colorado State University, who attended to the myriad details required for the smooth and successful operation of the workshop during both the 1992 session at Manhattan College, and the 1993 session at Colorado State. Among lecturers who assisted in the presentation of course material during the 1992 session were: Anthony J. Buonicore, president, Environmental Data Resources, Inc.; Dr. Marvin Fleishman, University of Louisville; Dr. Michael J. Hyland, Snamprogetti, Inc.; Dr. Paul Marnell, Manhattan College; Young McGuinn, Merck, Sharp and Dohme, Inc.; Ann Marie Nista and Peter F. Schmidt, Malcolm Pirnie, Inc.; Dr. Stuart Slater, Manhattan College; Jim Stouch, Malcolm Pirnie, Inc.; Valintin Tarasenko, Pfizer, Inc.; and Dr. John Wilcox, Manhattan College.

The editors also wish to acknowledge the National Science Foundation, without whose support this book would not have been possible.

It is the hope of the editors and contributors of this workbook that *Pollution Prevention: Problems and Solutions* will provide needed support for those faculty developing courses in P^2, and that it will become a useful resource for the training of engineers and scientists in the understanding of this critical topic.

CONTRIBUTORS

Nandkumar Bakshani, Department of Chemical Engineering, University of California, Los Angeles

Howard Beim, Department of Mathematics & Science, U.S. Merchant Marine Academy, Kings Point, New York

R. Ryan Dupont, Department of Civil and Environmental Engineering, Utah State University, Logan

Ihab H. Farag, Department of Chemical Engineering, University of New Hampshire, Durham

Kumar Ganesan, Department of Environmental Engineering, Montana Tech, Butte

Gary L. Hickernell, Division of Natural Sciences, Keuka College, Keuka Park, New York

Sonia M. Kreidenweis, Department of Atmospheric Science, Colorado State University, Fort Collins

Reid Lea, Department of Civil Engineering, Louisiana State University, Baton Rouge

Jennaro Maffia, Department of Chemical Engineering, Widener University, Chester, Pennsylvania

Daniel E. Medina, Department of Civil Engineering, Northeastern University, Boston, Massachusetts

Angelos Protopapas, Department of Civil and Environmental Engineering, Polytechnic University
Brooklyn, New York

Lisa A. Riedl, Department of Civil Engineering, University of Wisconsin-Platteville

Christian Roy, Department of Chemical Engineering, Laval University, Quebec, Canada

Dilip K. Singh, Department of Chemical Engineering, Youngstown State University, Ohio

Javad Tavakoli, Department of Chemical Engineering, Lafayette College, Easton, Pennsylvania

Jay R. Turner, Department of Chemical Engineering, Washington University, St. Louis, Missouri

Dean L. Ulrichson, Department of Chemical Engineering, Iowa State University, Ames

Note: Each problem statement (Chapter 2) includes the initials of the contributor who prepared the problem. For example, "[smk]" in a problem statement indicates that Sonia M. Kreidenweis prepared that problem. These initials appear in brackets immediately preceding the problem statement text.

CHAPTER 1: POLLUTION PREVENTION OVERVIEW
by
Brent Wainwright & Louis Theodore

INTRODUCTION

The amount of waste generated in the United States has reached staggering proportions; according to the United States Environmental Protection Agency (EPA), 250 million tons of solid waste alone are generated annually. Although both the Resource Conservation and Recovery Act (RCRA) and the Hazardous and Solid Waste Act (HSWA) encourage businesses to minimize the wastes they generate, the majority of our current environmental protection efforts are centered around treatment and pollution clean-up.

The passage of the Pollution Prevention Act of 1990 has redirected industry's approach to environmental management; pollution prevention has now become the environmental option of this decade and the 21st century. Whereas typical waste management strategies concentrate on "end-of-pipe" pollution control, pollution prevention attempts to handle waste at the source (i.e., source reduction). As waste handling and disposal costs increase, the application of pollution prevention measures is becoming more attractive than ever before. Industry is currently exploring the advantages of multimedia waste reduction and developing agendas to strengthen environmental design while lessening production costs.

There are profound opportunities for both the individual and industry to prevent the generation of waste; indeed, pollution prevention is today primarily stimulated by economics, legislation, liability concerns, and the enhanced environmental benefit of managing waste at the source. The EPA's Pollution Prevention Act of 1990 has established pollution prevention as a national policy declaring "waste should be prevented or reduced at the source wherever feasible, while pollution that cannot be prevented should be recycled in an environmentally safe manner" (U.S. EPA, 1991a). The EPA's policy establishes the following hierarchy of waste management:

1. Source Reduction; 2. Recycling/Reuse; 3. Treatment; 4. Ultimate Disposal

The hierarchy's categories are prioritized so as to promote the examination of each individual alternative prior to the investigation of subsequent options (i.e., the most preferable alternative should be thoroughly evaluated before consideration is given to a less accepted option.) Practices which decrease, avoid, or eliminate the generation of waste are considered source reduction and can include the implementation of procedures as simple and economical as good house-keeping. Recycling is the use, reuse or reclamation of wastes and/or materials which may involve the incorporation of waste recovery techniques (e.g., distillation, filtration). Recycling can be performed at the facility (i.e., on-site) or at an off-site reclamation facility. Treatment involves the destruction or detoxification of wastes into nontoxic or less toxic materials by chemical, biological or physical methods, or any combination of these control methods. Disposal has been included in the hierarchy because it is recognized that residual wastes will exist; the EPA's so called "ultimate disposal" options include land-filling, land farming, ocean dumping and deep-well injection. However, the term "ultimate disposal" is a misnomer, but is included here because of its adoption by the EPA. Table I provides a rough timetable demonstrating our national approach to waste management. Note how waste management has begun to shift from pollution control to pollution prevention.

TABLE I - Waste Management Timetable

Prior to 1945	No Control
1945 - 1960	Little Control
1960 - 1970	Some Control
1970 - 1975	Greater Control (EPA Founded)
1975 - 1980	More Sophisticated Control
1980 - 1985	Reasonably Available Control Technology (RACT)
1985 - 1990	Best Available Control Technology (BACT)
1990 - 1995	Pollution Prevention Act
	Lowest Achievable Emission Rate (LAER)
	Maximum Achievable Control Technology (MACT)
>1995	???

The development of waste management practices in the United States has recently moved toward securing a new pollution prevention ethic. The performance of pollution prevention assessments and their subsequent implementation will encourage increased research into methods that will further aid in the reduction of hazardous wastes.

One of the most important and propitious consequences of the pollution prevention movement will be the development of standardized life-cycle cost accounting procedures. "Life-cycle" is a perspective which considers the true costs of product production and/or services provided and utilized by analyzing the price associated with potential environmental degradation and energy consumption, as well as more customary costs like capital expenditure and operating expenses.

The remainder of this text will be concerned with providing the reader with the necessary background to understand the meaning of pollution prevention and its useful implementation. Assessment procedures and the economic benefits derived from managing pollution at the source are discussed along with methods of cost-accounting for pollution prevention. Additionally, regulatory and non-regulatory methods to promote pollution prevention and overcome barriers are examined, and ethical considerations presented. By eliminating waste at the source, we can all participate in the protection of our environment by reducing the amount of waste material that would otherwise need to be treated or ultimately disposed; accordingly, attention is also given to pollution prevention in both the domestic and business office environments.

POLLUTION PREVENTION HIERARCHY

As discussed in the Introduction, the hierarchy set forth by the USEPA in the Pollution Prevention Act establishes an order to which waste management activities should be employed to reduce the quantity of waste generated. The preferred method is source reduction, as indicated in Figure 1. This approach actually precedes traditional waste management by addressing the source of the problem prior to its occurrence.

Although the EPA's policy does not consider recycling or treatment as actual pollution prevention methods per se, these methods present an opportunity to reduce the amount of waste that might otherwise be discharged into the environment. Clearly, the definition of pollution prevention and its synonyms (e.g., waste minimization) must be understood to fully appreciate and apply these techniques.

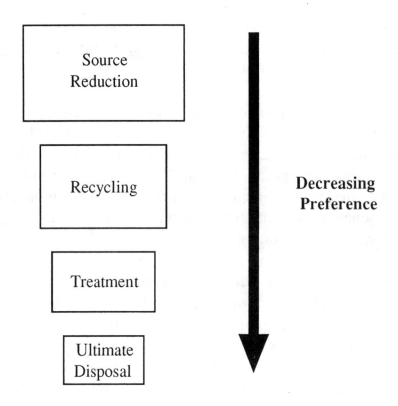

Figure 1.   Pollution prevention hierarchy (U.S. EPA, 1991a).

Waste minimization generally considers all of the methods in the EPA hierarchy (except for disposal) appropriate to reduce the volume or quantity of waste requiring disposal (i.e., source reduction). The definition of source reduction as applied in the Pollution Prevention Act, however, is "any practice which reduces the amount of any hazardous substance, pollutant or contaminant entering any waste stream or otherwise released into the environment...prior to recycling, treatment or disposal" (U.S. EPA, 1991a).   Source reduction reduces the amount of waste generated; it is therefore considered true pollution prevention and has the highest priority in the EPA hierarchy.

Recycling (reuse, reclamation) refers to the use or reuse of materials that would otherwise be disposed of or treated as a waste product. Wastes that cannot be directly reused may often be recovered on-site through methods such as distillation. When on-site recovery or reuse is not feasible due to quality specifications or the inability to perform recovery on-site, off-site recovery at a permitted commercial recovery facility is often a possibility. Such management techniques are considered secondary to source reduction and should only be used when pollution cannot be prevented.

The treatment of waste is the third element of the hierarchy and should be utilized only in the absence of feasible source reduction or recycling opportunities. Waste treatment involves the use of chemical, biological, or physical processes to reduce or eliminate waste material. The incineration of wastes is included in this category and is considered "preferable to other treatment methods (i.e., chemical, biological, and physical) because incineration can permanently destroy the hazardous components in waste materials" (Theodore and McGuinn, 1992).

Of course, many of these elements are used by industry in combination to achieve the greatest waste reduction. Residual wastes which cannot be prevented or otherwise managed are then disposed of only as a last resort.

Figure 2 provides a schematic representation of the two preferred pollution prevention techniques (i.e., source reduction and recycling).

MULTIMEDIA ANALYSIS AND LIFE CYCLE COST ANALYSIS

Multimedia Analysis

In order to properly design and then implement a pollution prevention program, sources of all wastes must be fully understood and evaluated. A multimedia analysis involves a multifaceted approach. It must not only consider one waste stream but all potentially contaminant media (e.g., air, water, land). Our past waste management practices have been concerned primarily with treatment. All too often, such methods solve one waste problem by transferring a contaminant from one medium to another (e.g., air-stripping); such waste shifting is not pollution prevention or waste reduction.

Pollution prevention techniques must be evaluated through a thorough consideration of all media, hence the term "multimedia". This approach is a clear departure from previous pollution treatment or control techniques where it was acceptable to transfer a pollutant from one source to another in order to solve a waste problem. Such strategies merely provide short-term solutions to an ever increasing problem. As an example, air pollution control equipment prevents or reduces the discharge of waste into the air but at the same time can produce a solid hazardous waste problem.

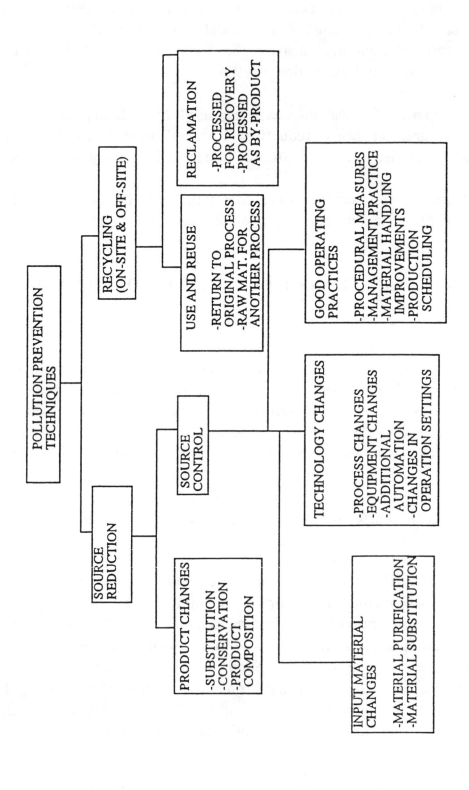

Figure 2.   Pollution prevention techniques.

Life-Cycle Analysis

The aforementioned multimedia approach to evaluating a product's waste stream(s) aims to ensure that the treatment of one waste stream does not result in the generation or increase of an additional waste output. Clearly, impacts resulting during the production of a product must be evaluated over its entire history or "life-cycle."

A life-cycle analysis, or "Total Systems Approach," (Theodore, 1993) is crucial to identifying opportunities for improvement. This type of evaluation identifies "energy use, material inputs, and wastes generated during a product's life: from extraction and processing of raw materials, to manufacture and transport of a product to the marketplace, and, finally, to use and dispose of the product" (World Wildlife Fund, 1991).

During a forum convened by the World Wildlife Fund and the Conservation Foundation in May 1990, various steering committees recommended that a three-part life-cycle model be adopted. This model consists of the following:

- An inventory of materials and energy used, and environmental releases from all stages in the life of a product or process;
- An analysis of potential environmental effects related to energy use and material resources and environmental releases; and,
- An analysis of the changes needed to bring about environmental improvements for the product or process under evaluation.

Traditional cost analyses often fail to include factors relevant to future damage claims resulting from litigation, the depletion of natural resources, the effects of energy use, etc. As such, waste management options such as treatment and disposal may appear preferential if an overall life-cycle cost analysis is not performed. It is evident that environmental costs from "cradle-to-grave" have to be evaluated together with more conventional production costs to accurately ascertain genuine production costs. In the future, a total systems approach will most likely involve a more careful evaluation of pollution, energy, and safety issues. For example, if one was to compare the benefits of coal versus oil as a fuel source for an electric power plant, the use of coal might be considered economically favorable. In addition to the cost issues, however, one must be concerned with the environmental effects of coal mining, transportation and storage prior to use as a fuel. We have a tendency to overlook the fact that there are serious health and safety matters (e.g., miner exposure) which must be considered, along with the effects of fugitive emissions. When these effects are weighed alongside of standard economic factors, the cost benefits of coal usage may no longer appear valid. Thus, many of the economic benefits associated with pollution prevention are often unrecognized due to inappropriate cost accounting methods. For this reason, economic considerations are detailed later.

POLLUTION PREVENTION ASSESSMENT PROCEDURES

The first step in establishing a pollution prevention program is the obtainment of management commitment. Management commitment is necessary given the inherent need for project structure and control. Management will determine the amount of funding allotted for the program as well as specific program goals. The data collected during the actual evaluation are then used to develop options for reducing the types and amounts of waste generated. Figure 3 depicts a systematic approach that can be used during the procedure. After a particular waste stream or area of concern is identified, feasibility studies are performed involving both economic and technical considerations. Finally, preferred alternatives are implemented. The four phases of the assessment (i.e., planning and organization, assessment, feasibility, and implementation) are introduced in the following subsections. Sources of additional information, as well as information on industrial programs, is also provided in this section.

Planning and Organization

The purpose of this phase is to obtain management commitment, define and develop program goals, and, to assemble a project team. Proper planning and organization are crucial to the successful performance of the pollution prevention assessment. Both managers and facility staff play important roles in the assessment procedure by providing the necessary commitment and familiarity with the facility, its processes, and current waste management operations. It is the benefits of the program, including economic advantages, liability reduction, regulatory compliance and improved public image, etc., that often leads to management support.

Once management has made a commitment to the program and goals have been set, a program task force is established. The selection of a team

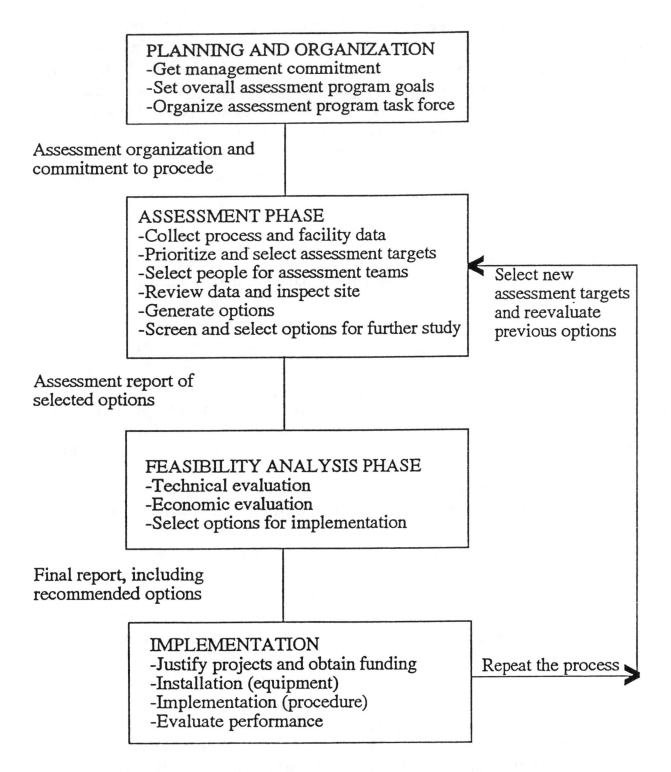

Figure 3. Pollution Prevention Assessment Procedures.

leader will be dependent upon many factors including their ability to effectively interface with both the assessment team and management staff.

The task force must be capable of identifying pollution reduction alternatives, as well as be cognizant of inherent obstacles to the process. Barriers frequently arise from the anxiety associated with the belief that the program will negatively affect product quality or result in production losses. According to an EPA survey, 30 percent of the respondents were concerned that product quality would decline if waste minimization techniques were implemented (U.S. EPA, 1991b). As such, the assessment team, and the team leader in particular, must be ready to react to these and other concerns (Theodore, 1993).

Assessment Phase

The assessment phase aims to collect data needed to identify and analyze pollution prevention opportunities. Assessment of the facility's waste reduction needs includes the examination of hazardous waste streams, process operations, and the identification of techniques that often promise the reduction of waste generation. Information is often derived from observations made during a facility walk-through, interviews with employees (e.g., operators, line workers), and review of site or regulatory records. The American Society of Testing and Materials (ASTM) suggests the following information sources be reviewed, as available (ASTM, 1992):

- Product design criteria.
- Process flow diagrams for all solid waste, wastewater, and air emissions sources.
- Site maps showing the location of all pertinent units (e.g., pollution control devices, points of discharge).
- Environmental documentation, including: Material Safety Data Sheets (MSDS), military specification data, permits (e.g., NPDES, POTW, RCRA), SARA Title III reports, waste manifests, and any pending permits or application information.
- Economic data, including: cost of raw material management; cost of air, wastewater, and hazardous waste treatment; waste management operating and maintenance costs; and waste disposal costs.
- Managerial information: environmental policies and procedures; prioritization of waste management concerns; automated or computerized waste management systems; inventory and distribution procedures; maintenance scheduling practices; planned

modifications or revisions to existing operations that would impact waste generation activities; and the basis of source reduction decisions and policies.

The use of process flow diagrams and material balances are worthwhile methods to "quantify losses or emissions, and provide essential data to estimate the size and cost of additional equipment, data to evaluate economic performance, and a baseline for tracking the progress of minimization efforts" (Theodore, 1993). Material balances should be applied to individual waste streams or processes and then utilized to construct an overall balance for the facility. Details on these calculations are available in the literature (Thodore and Allen, 1993). In addition, an introduction to this subject is provided in the next section.

The data collected are then used to prioritize waste streams and operations for assessment. Each waste stream is assigned a priority based on corporate pollution prevention goals and objectives. Once waste origins are identified and ranked, potential methods to reduce the waste stream are evaluated. The identification of alternatives is generally based on discussions with the facility staff, review of technical literature, and contacts with suppliers, trade organizations and regulatory agencies.

Alternatives identified during this phase of the assessment are evaluated using screening procedures so as to reduce the number of alternatives requiring further exploration during the feasibility analysis phase. The criteria used during this screening procedure include: cost-effectiveness; implementation time; economic, compliance, safety and liability concerns; waste reduction potential; and whether the technology is proven (Theodore, 1993; Thodore and Allen, 1993). Options which meet established criteria are then examined further during the feasibility analysis.

Feasibility Analysis

Preferred alternative selection is performed by an evaluation of technical and economic considerations. The technical evaluation determines whether or not a given option will work as planned. Some typical considerations follow:

- Safety concerns
- Product quality impacts or production delays during implementation
- Labor and/or training requirements
- Creation of new environmental concerns
- Waste reduction potential
- Utility and budget requirements
- Space and compatibility concerns

If an option proves to be technically ineffective or inappropriate, it is deleted from the list of potential alternatives. Either following or concurrent with the technical evaluation, an economic study is performed weighing standard measures of profitability such as payback period, investment returns, and net present value. Many of these costs (or more appropriately, cost savings) may be substantial yet are difficult to quantify. (Refer to *Economic Considerations Associated with Pollution Prevention*).

Implementation

The findings of the overall assessment are used to demonstrate the technical and economic worthiness of program implementation. Once appropriate funding is obtained, the program is implemented not unlike any other project requiring new procedures or equipment. When preferred waste pollution prevention techniques are identified, they are implemented, and should become part of the facility's day-to-day management and operation. Subsequent to the program's execution, its performance should be evaluated in order to demonstrate effectiveness, generate data to further refine and augment waste reduction procedures, and maintain management support.

It should be noted that waste reduction, energy conservation, and safety issues are interrelated and often complementary to each other. For example, the reduction in the amount of energy a facility consumes results in reduced emissions associated with the generation of power. Energy expenditures associated with the treatment and transport of waste are similarly reduced when the amount of waste generated is lessened; at the same time worker safety is elevated due to reduced exposure to hazardous materials.

Sources of Information

The successful development and implementation of any pollution prevention program are not only dependent on a thorough understanding of the facility's operations but also require an intimate knowledge of current opportunities and advances in the field. In fact, 32 percent of industry respondents to an EPA survey identified the lack of technical information as a major factor delaying or preventing the implementation of a waste minimization program (U.S. EPA, 1991b). Fortunately, the EPA has developed a national Pollution Prevention Information Clearinghouse (PPIC) and the Pollution Prevention Information Exchange System (PIES) to facilitate the exchange of information needed to promote pollution prevention through efficient information transfer (U.S. EPA, 1991c).

PPIC is operated by the EPA's Office of Research and Development and the Office of Pollution Prevention. The clearinghouse is comprised of four elements:

- *Repository* - including a hard copy reference library and collection center and an on-line information retrieval and ordering system.
- *PIES* - a computerized conduit to data bases and document ordering, accessible via modem and personal computer - (703) 506-1025.
- *Hotline* - PPIC uses the RCRA/Superfund and Small Business Ombudsman Hotlines as well as a PPIC technical assistance line to answer pollution prevention questions, access information in the PPIC, and assist in document ordering and searches. To access PPIC by telephone, call:

    RCRA/Superfund Hotline (800) 242-9346
    Small Business Ombudsman Hotline (800) 368-5888
    PPIC Technical Assistance (703) 821-4800

- *Networking and Outreach* - PPIC compiles and disseminates information packets and bulletins, and initiates networking efforts with other national and international organizations.

Additionally, the EPA publishes a newsletter entitled, *Pollution Prevention News*, which contains information including EPA news, technologies, program updates, and case studies. The EPA's Risk Reduction Engineering Laboratory and the Center for Environmental Research Information has published several guidance documents developed in cooperation with the California Department of Health Services. The manual's supplement generic waste reduction information is presented in the EPA's *Waste Minimization Opportunity Assessment Manual* (U.S. EPA, 1988). Additional information is available through PPIC.

Pollution prevention or waste minimization programs have been established at the State level and as such are good sources of information. Both Federal and State agencies are working with universities and research centers and may also provide assistance. For example, the American Institute of Chemical Engineers has established the Center for Waste Reduction Technologies (CWRT), a program based on targeted research, technology transfer, and enhanced education.

Industry Programs

A significant pollution prevention resource may very well be found with the "competition." Several large companies have established well-known programs that have successfully incorporated pollution prevention practices into their manufacturing processes. These include, but are not limited to: 3M - *Pollution Prevention Pays* (3P); Dow Chemical - *Waste Reduction Always Pays* (WRAP); Chevron - *Save Money And Reduce Toxics* (SMART); and, the General Dynamics - *Zero Discharge Program*.

Smaller companies can benefit by the assistance offered by these larger corporations. It is clear that access to information is of major importance when implementing efficient pollution prevention programs. By adopting such programs, industry is affirming pollution prevention's application as a good business practice and not simply a "noble" effort.

ASSESSMENT PHASE MATERIAL BALANCE CALCULATIONS
(Theodore, 1993; Theodore and Reynolds, 1989)

One of the key elements of the assessment phase of a pollution prevention program involves mass balance equations. These calculations are often referred to as material balances; the calculations are performed via the conservation law for mass. The details of this often-used law are described below.

The conservation law for mass can be applied to any process or system. The general form of the law follows:

mass in - mass out + mass generated = mass accumulated

This equation can be applied to the total mass involved in a process or to a particular species, on either a mole or mass basis. The conservation law for mass can be applied to steady-state or unsteady-state processes and to batch or continuous systems. A steady-state system is one in which there is no change in conditions (e.g., temperature, pressure) or rates of flow with time at any given point in the system; the "accumulation" term then becomes zero. If there is no chemical reaction, the "generation" term is zero. All other processes are classified as unsteady-state.

To isolate a system for study, the system is separated from the surroundings by a boundary or envelope which may either be real (e.g., a reactor vessel) or imaginary. Mass crossing the boundary and entering the system is part of the "mass-in" term. The equation may be used for any compound whose quantity does not change by chemical reaction, or for any chemical element, regardless of whether it has participated in a chemical reaction. Furthermore, it may be written for one piece of equipment, several pieces of equipment, or around an entire process (i.e., a total material balance).

The conservation of mass law finds a major application during the performance of pollution prevention assessments. As described earlier, a pollution prevention assessment is a systematic, planned procedure with the objective of identifying methods to reduce or eliminate waste. The assessment process should characterize the selected waste streams and processes (ICF Technology Incorporated, 1989) - a necessary ingredient if a material balance is to be performed. Some of the data required for the material balance calculation may be collected during the first review of site-specific data; however, in some instances, the information may not be collected until an actual site walk-through is performed.

Simplified mass balances should be developed for each of the important waste-generating operations to identify sources and gain a better understanding of the origins of each waste stream. Since a mass balance is essentially a check to make sure that what goes into a process (i.e., the total mass of all raw materials), what leaves the process (i.e., the total mass of the product(s) and by-products), the material balance should be made individually for all components that enter and leave the process. When chemical reactions take place in a system, there is an advantage to doing "elemental balances" for specific chemical elements in a system. Material balances can assist in determining concentrations of waste constituents where analytical test data are limited. They are particularly useful when there are points in the production process where it is difficult or uneconomical to collect analytical data.

Mass balance calculations are particularly useful for quantifying fugitive emissions, such as evaporative losses. Waste stream data and mass balances will enable one to track flow and characteristics of the waste streams over time. Since in most cases the accumulation equals zero (steady-state operation), it can then be assumed that any build-up is actually leaving the process through fugitive emissions or other means. This will be useful in identifying trends in waste/pollutant generation and will also be critical in the task of measuring the performance of implemented pollution prevention options. The result of these activities is a catalog of waste streams that provides a description of each waste, including quantities, frequency of discharge, composition and other important information useful for material balance. Of course, some assumptions or educated estimates will be needed when it is impossible to obtain specific information.

By performing a material balance in conjunction with a pollution prevention assessment, the amount of waste generated becomes known. The success of the pollution prevention program can therefore be measured by using this information on baseline generation rates (i.e., that rate at which waste is generated without pollution prevention considerations).

BARRIERS AND INCENTIVES TO POLLUTION PREVENTION

As discussed previously, industry is beginning to realize that there are profound benefits associated with pollution prevention including cost effectiveness, reduced liability, enhanced public image, and regulatory compliance. Nevertheless, there are barriers or disincentives identified with pollution prevention. This section will briefly outline both barriers and incentives which may need to be confronted or considered during the evaluation of a pollution prevention program.

Barriers to Pollution Prevention ("The Dirty Dozen")

There are numerous reasons why more businesses are not reducing the wastes they generate. The following "dirty dozen" are common disincentives:

1. *Technical Limitations* - Given the complexity of present manufacturing processes, waste streams exist that cannot be reduced with current technology. The need for continued research and development is evident.

2. *Lack of Information* - In some instances, the information needed to make a pollution prevention decision may be confidential or is difficult to obtain. In addition, many decision makers are simply unaware of the potential opportunities available regarding information to aid in the implementation of a pollution prevention program.

3. *Consumer Preference Obstacles* - Consumer preference strongly affects the manner in which a product is produced, packaged and marketed. If the implementation of a pollution prevention program results in the increase in the cost of a product, or decreased convenience or availability, consumers might be reluctant to use it.

4. *Concern over Product Quality Decline* - The use of a less hazardous material in a product's manufacturing process may result in decreased life, durability, or competitiveness.

5. *Economic Concerns* - Many companies are unaware of the economic advantages associated with pollution prevention. Legitimate concerns may include decreased profit margins or the lack of funds required for the initial capital investment.

6. *Resistance to Change* - The unwillingness of many businesses to change is rooted in their reluctance to try technologies which may be unproven, or based on a combination of the barriers discussed in this section.

7. *Regulatory Barriers* - Existing regulations that have created incentives for the control and containment of wastes are at the same time discouraging the exploration of pollution prevention alternatives. Moreover, since regulatory enforcement is often intermittent, current legislation can weaken waste reduction incentives.

8. *Lack of Markets* - The implementation of pollution prevention processes and the production of environmentally-friendly products will be of no avail if markets do not exist for such goods. As an

example, the recycling of newspaper in the United States has resulted in an overabundance of waste paper without markets prepared to take advantage of this "raw" material.

9. *Management Apathy* - Many managers capable of making decisions to begin pollution prevention activities do not realize the potential benefits of pollution prevention and may therefore take on a attitude of passiveness.

10. *Institutional Barriers* - In an organization without a strong infrastructure to support pollution prevention plans, waste reduction programs will be difficult to implement. Similarly, if there is no mechanism in-place to hold individuals accountable for their actions, the successful implementation of a pollution prevention program will be limited.

11. *Lack of Awareness of Pollution Prevention Advantages* - As mentioned in *economic concerns*, decision makers may merely be uninformed of the benefits associated with pollution reduction.

12. *Concern over the Dissemination of Confidential Product Information* - If a pollution prevention assessment reveals confidential data pertinent to a company's product, fear may exist that the organization will lose a competitive edge with other businesses in the industry.

Pollution Prevention Incentives ("A Baker's Dozen")

Various means exist to encourage pollution prevention through regulatory measures, economic incentives, and technical assistance programs. Since the benefits of pollution prevention undoubtedly surpass prevention barriers, a "baker's dozen" incentives list is presented below:

1. *Economic Benefits* - The most obvious economic benefits associated with pollution prevention are the savings which result from the elimination of waste storage, treatment, handling, transport, and disposal. Additionally, less tangible economic benefits are realized in terms of decreased liability, regulatory compliance costs (e.g., permits), legal and insurance costs, and improved process efficiency. Pollution prevention almost always pays for itself, particularly when the time investment required to comply with regulatory standards is considered. Several of these economic benefits are discussed separately below.

2. *Regulatory Compliance* - Quite simply, when wastes are not generated, compliance issues are not a concern. Waste management costs associated with record-keeping, reporting, and laboratory

analysis are reduced or eliminated. Pollution prevention's proactive approach to waste management will better prepare industry for the future regulation of many hazardous substances and wastes which are currently unregulated. Regulations have, and will continue to be, a moving target.

3. *Liability Reduction* - Facilities are responsible for their wastes from "cradle-to-grave." By eliminating or reducing waste generation, future liabilities can also be decreased. Additionally, the need for expensive pollution liability insurance requirements may be abated.

4. *Enhanced Public Image* - Consumers are interested in purchasing goods that are safer for the environment and this demand, depending on how they respond, can mean success or failure for many companies. Business should therefore be sensitive to consumer demands and use pollution prevention efforts to their utmost advantage by producing goods that are environmentally friendly.

5. *Federal and State Grants* - Federal and State grant programs have been developed to strengthen pollution prevention programs initiated by states and private entities. The EPA's Pollution Prevention By and For Small Business Grant Program awards grants to small businesses to assist their development and demonstration of new pollution prevention technologies.

6. *Market Incentives* - Public demand for environmentally-preferred products has generated a market for recycled goods and related products; products can be designed with these environmental characteristics in mind, offering a competitive advantage. In addition, many private and public agencies are beginning to stimulate the market for recycled goods by writing contracts and specifications which call for the use of recycled materials.

7. *Reduced Waste Treatment Costs* - As discussed in *economic benefits,* the increasing costs of traditional end-of-pipe waste management practices are avoided or reduced through the implementation of pollution prevention programs.

8. *Potential Tax Incentives* - As an effort to promote pollution prevention, taxes may eventually need to be levied to encourage waste generators to consider reduction programs. Conversely, tax breaks to corporations which utilize pollution prevention methods could similarly be developed to foster pollution prevention.

9. *Decreased Worker Exposure* - By reducing or eliminating chemical exposures, businesses benefit by lessening the potential for chronic workplace exposure, and serious accidents and emergencies. The burden of medical monitoring programs, personal exposure

monitoring, and potential damage claims are also reduced.

10. *Decreased Energy Consumption* - As mentioned previously, energy conservation and pollution prevention are often interrelated and complementary to each other. Energy expenditures associated with the treatment and transport of waste are reduced when the amount of waste generated is lessened while at the same time the pollution associated with energy consumed by these activities is abated.

11. *Increased Operating Efficiencies* - A potential beneficial side effect of pollution prevention activities is a concurrent increase in operating efficiency. Through a pollution prevention assessment, the assessment team can identify sources of waste that result in hazardous waste generation <u>and</u> loss in process performance. The implementation of a reduction program will often rectify such problems through modernization, innovation, and the implementation of good operating practices.

12. *Competitive Advantages* - By taking advantage of the many benefits associated with pollution prevention, businesses can gain a competitive edge.

13. *Reduced Negative Environmental Impacts* - Through an evaluation of pollution prevention alternatives which consider a total systems approach, consideration is given to the negative impact of environmental damage to natural resources and species which occurs during raw material procurement and waste disposal. The performance of pollution prevention endeavors will therefore result in enhanced environmental protection.

It is evident that the majority of the obstacles to pollution prevention are based on either a lack of information or an anxiety associated with economic concerns. By strengthening the exchange of information among businesses, a better understanding of the unique benefits of pollution prevention will be realized. The development of new markets by means of regulatory and economic incentives will further assist the effective implementation of waste reduction. Additionally, the development of accurate life-cycle cost accounting methods will be instrumental in discerning the true financial advantages of pollution prevention.

Various combinations of the pollution prevention barriers provided above have appeared on numerous occasions in the literature, and in many different forms. However, there is one other concern that both industry and the taxpayer should be aware of. Carol Browner, EPA Administrator, has repeatedly claimed that pollution prevention is the organization's top priority. Nothing could be further from the truth. Despite near unlimited resources, the EPA has contributed little to furthering the pollution prevention effort. The EPA offices in Washington, Research Park Triangle, and Region II have exhibited a level of bureaucratic indifference that has surpassed even the traditional attitudes of many EPA employees. It is virtually impossible to contact any responsible pollution prevention individual at the EPA. Calls are rarely returned. Letters are rarely returned. On the rare occasion when contact is made, the individual typically passes the caller onto someone else who "really is in a better position to help you," and the cycle starts all over again (Theodore, 1994).

This standard bureaucratic phenomena has been experienced by others in industry and the EPA (Theodore, 1994). Two letters of complaint to the EPA Region II Administrator resulted in a response that was somewhat cynical and suggestive of a reprimand. Ms. Browner chose not to reply to the complaints (Theodore, 1994). Notwithstanding some of the above comments, pollution prevention efforts have been successful in industry because these programs have often either produced profits or reduced costs, or both. The driving force for these successes has primarily been economics and not the EPA.

ECONOMIC CONSIDERATIONS ASSOCIATED WITH POLLUTION PREVENTION PROGRAMS

The purpose of this section is to outline the basic elements of a pollution prevention cost accounting system which incorporates both traditional and less tangible economic variables. The intent is not to present a detailed discussion of economic analysis but to help identify the more important elements that must be considered to properly quantify pollution prevention options.

The greatest driving force behind any pollution prevention plan is the promise of economic opportunities and cost savings over the long-term. Pollution prevention is now recognized as one of the lowest-cost options for waste/pollutant management. Hence, an understanding of the economics involved in pollution prevention programs/options is quite important in making decisions at both the engineering and management levels. Every engineer should be able to execute an economic evaluation of a proposed project. If the project cannot be justified economically after all factors (these will be discussed in more detail below) and considerations have been taken into account, it should obviously not be pursued. The earlier such a project is identified, the fewer resources will be wasted.

Before the true cost or profit of a pollution prevention program can be evaluated, the factors contributing to the economics must be recognized. There are two traditional contributing factors: capital costs and operating costs, but there are other important costs and benefits associated with pollution prevention that need to be quantified if a meaningful economic analysis is going to be performed. Table II demonstrates the evolution of various cost accounting methods. Although Tables I (see Introduction section) and II are not directly related, the reader is left with the option of comparing some of the similarities between the two.

The Total Systems Approach (TSA) referenced in Table II aims to quantify not only the economic aspects of pollution prevention but also the social costs associated with the production of a product or service from cradle-to-grave (i.e., life-cycle). The TSA attempts to quantify less tangible benefits such as the reduced risk derived from not using a hazardous substance. The future is certain to see more emphasis placed on the TSA approach in any pollution prevention program. As described earlier, a utility considering the option of converting from a gas-fired

TABLE II - Economic Analysis

Prior to 1945	Capital Costs Only
1945 - 1960	Capital and Some Operating Costs
1960 - 1970	Capital and Operating Costs
1970 - 1975	Capital, Operating, and Some Environmental Control Costs
1975 - 1980	Capital, Operating, and Environmental Control Costs
1980 - 1985	Capital, Operating, and More Sophisticated Environmental Control Costs
1985 - 1990	Capital, Operatings, and Environmental Controls, and some Life-Cycle Analysis (Total Systems Approach)
1990 - 1995	Capital, Operating, and Environmental Control Costs and Life-Cycle Analysis (Total Systems Approach)
>1995	???

boiler to coal-firing is today not concerned with the environmental effects and implications associated with mining, transporting, storing, etc., the coal prior to its usage as an energy feedstock. Pollution prevention approaches in the mid-to-late 1990s will become more aware of this need.

Figure 4 depicts the interrelation between the various factors that together make-up the TSA. It is the outcome of this type of analysis that should determine the ultimate fate of a pollution prevention project. Once the total costs of the pollution prevention "change" have been estimated, a determination as to whether or not the project will be profitable can be made. This often involves converting all economic contributions to an annualized basis (e.g., annualized capital costs or "ACC"). If more than one pollution prevention project is under study, this method provides a basis for comparing alternate proposals and for choosing the best alternative.

The economic evaluation referred to above is usually carried out using standard measures of profitability. Each company and organization has its own economic criteria for selecting projects for implementation. In performing an economic evaluation, various costs and savings must be considered. The economic analysis presented in this section represents a preliminary, rather than a detailed, analysis. For smaller facilities with only a few (and perhaps simple) processes, the entire pollution prevention assessment procedure will tend to be much less formal. In this situation, several obvious pollution prevention options, such as the installation of flow controls and good operating practices, may be implemented with little or no economic evaluation. In these instances, no complicated analyses are necessary to demonstrate the advantages of adopting the selected pollution prevention option. A proper perspective must also be maintained between the magnitude of savings that a potential option may offer and the amount of manpower required to do the technical and economic feasibility analyses. A short description of the various economic factors - including capital and operating costs, and "other" considerations follows.

Once identified, the costs and/or savings are placed into their appropriate categories and quantified for subsequent analysis. Equipment cost is a function of many variables, one of the most significant of which is capacity. Other important variables include operating temperature and/or pressure conditions, and degree of equipment sophistication. Preliminary estimates are often made using simple cost-capacity

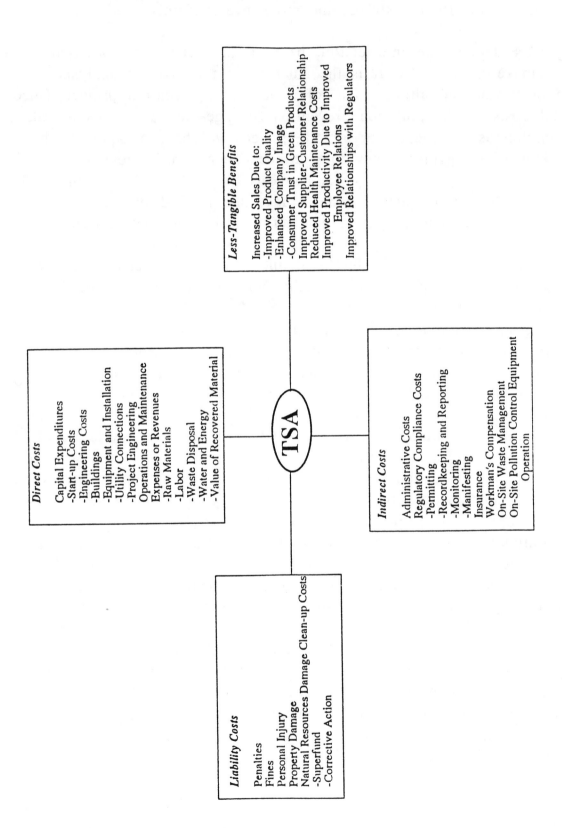

Figure 4.   Total Systems Approach (TSA) cost inventory.

relationships that are valid when the other variables are confined to a narrow range of values.

The usual technique for determining the capital costs (i.e., total capital costs, which include equipment design, purchase, and installation) for the facility is based on the factored method of establishing direct and indirect installation costs as a function of the known equipment costs. This is basically a modified Lang method, whereby cost factors are applied to known equipment costs (Neveril, 1978; Vatavuk and Neveril, 1980). The first step is to obtain from vendors the purchase prices of the primary and auxiliary equipment. The total base price, designated by X, which should include instrumentation, control, taxes, freight costs, and so on, serves as the basis for estimating the direct and indirect installation costs. These costs are obtained by multiplying X by the cost factors, which are available in the literature (Neveril, 1978; Vatavuk and Neveril, 1980; Vogel and Martin, 1983a, 1983b, 1983c; Ulrich, 1984).

The second step is to estimate the direct installation costs by summing all the cost factors involved in the direct installation costs, which include piping, insulation, foundation and supports, etc. The sum of these factors is designated as the DCF (direct installation cost factor). The direct installation costs are then the product of the DCF and X. The third step consists of estimating the indirect installation costs; that is, all the cost factors for the indirect installation costs (engineering and supervision, start-up, construction fees, etc.) are added; the sum is designated by ICF (indirect installation cost factor). The indirect installation costs are then the product of ICF and X. Once the direct and indirect installation costs have been calculated, the total capital cost (TCC) may be evaluated as follows:

$$TCC = X + (DCF)(X) + (ICF)(X)$$

This is then converted to annualized capital costs (ACC) with the use of the Capital Recovery Factor (CRF) which can be calculated from the following equation:

$$CRF = [(i)(1 + i)^n]/[(1 + i)^n - 1]$$

where n = projected lifetime of the project (yr), and i = annual interest rate (expressed as a fraction).

The annualized capital cost (ACC) is the product of the CRF and TCC and represents the total installed equipment cost distributed over the lifetime of the project. The ACC reflects the cost associated with the initial capital outlay over the depreciable life of the system. Although investment and operating costs can be accounted for in other ways, such as present-worth analysis, the capital recovery method is preferred because of its simplicity and versatility. This is especially true when comparing somewhat similar systems having different depreciable lives. In such decisions there are usually other considerations besides economic, but if all other factors are equal, the alternative with the lowest total annualized cost should be the most viable.

Operating costs can vary from site to site since these costs reflect local conditions (e.g., staffing practices, labor, utility costs). Operating costs, like capital costs, may be separated into two categories: direct and indirect costs. Direct costs are those that cover material and labor and are directly involved in operating the facility. These include labor, materials, maintenance and maintenance supplies, replacement parts, wastes, disposal fees, utilities, and laboratory costs. Indirect costs are those operating costs associated with, but not directly involved in, operating the facility; costs such as overhead (e.g., building-land leasing and office supplies), administrative fees, property taxes, and insurance fees fall into this category. However, the major direct operating costs are usually those associated with labor and materials.

The main problem with the traditional type of economic analysis is that it is difficult (in some cases impossible) to quantify some of the not-so-obvious economic merits of a pollution prevention program. Several considerations, in addition to those provided in Figure 4, have just recently surfaced as factors that need to be taken into account in any meaningful economic analysis of a pollution prevention effort. What follows is a summary of these considerations:

- Decreased long-term liabilities
- Regulatory compliance
- Regulatory recordkeeping
- Dealings with the EPA
- Dealings with state and local regulatory bodies
- Elimination or reduction of fines and penalties
- Potential tax benefits
- Customer relations
- Stockholder support (corporate image)
- Improved public image

- Reduced technical support
- Potential insurance costs and claims
- Effect on borrowing power
- Improved mental and physical well-being of employees
- Reduced health maintenance costs
- Employee morale
- Other process benefits
- Improved worker safety
- Avoidance of rising costs of waste treatment and/or disposal
- Reduced training costs
- Reduced emergency response planning

Many proposed pollution prevention programs have been squelched in their early stages because a comprehensive economic analysis was not performed. Until the effects described above are included, the true merits of a pollution prevention program may be clouded by incorrect and/or incomplete economic data. Can something be done by industry to remedy this problem? One approach is to use a modified version of the standard Delphi Panel that the authors of this work have modestly defined as the WTA (an acronym for the Wainwright-Theodore Approach). In order to estimate these "other" economic benefits of pollution prevention, several knowledgeable individuals within and perhaps outside the organization are asked to independently provide estimates, with explanatory details, on these economic benefits. Each individual in the panel is then allowed to independently review all responses. The cycle is then repeated until the group's responses approach convergence.

Finally, pollution prevention measures can provide a company with the opportunity of looking their neighbors in the eye and truthfully saying that all that can reasonably be done to prevent pollution is being done...in effect, the company is doing right by the environment. Is there an economic advantage to this? It is not only a difficult question to answer quantitatively but also a difficult one to answer. The reader is left with pondering the answer to this question.

OTHER ACCOUNTING CONSIDERATIONS

Through the use of a TSA or what the EPA refers to as a "Total Cost Assessment," a more correct financial picture of pollution prevention's financial benefits is obtained. A total cost assessment (TCA) is comprised of four elements: expanded cost inventory, extended time horizon, long-term financial indicators, and the direct allocation of costs to specific processes and products. These four components are briefly discussed below.

Expanded Cost Inventory

In order to perform an accurate financial analysis of any pollution prevention project, all of the costs and savings associated with the investment must be identified for consideration. Traditional budgeting techniques generally do not include all of the pollution prevention benefits because:

- They do not easily fit into standard categories (e.g., direct, equipment costs);
- They are difficult to quantify;
- A greater detail of analysis is often required to identify their costs or advantages; and,
- The benefits of a pollution prevention program need to be evaluated over longer time periods.

Many costs are obviously difficult to quantify or involve significant uncertainty. General Electric has developed a model to estimate risk and generate a cost per ton estimate for future environmentally liability (Zornberg and Wainright, 1993). This model attempts to predict the costs associated with "future claims for personal injury, economic losses, and natural resource damage, and future site remediation actions" (Zornberg and Wainright, 1993). In the absence of concrete cost quantifying methods, estimates may be made based on claims or penalties against a similar company, or, qualitatively characterized without a direct dollar value.

Expanded Time Horizon

Many of the less-tangible benefits of pollution prevention occur over a greater period of time than normally considered during the economic analysis of more conventional projects. As such, it is recommended that time horizons of five years or more be used to recognize the costs, savings, and revenues which develop during the long-term (U.S. EPA, 1992).

Long-Term Financial Indicators

Both the EPA and ASTM recommend the use of long-term financial indicators during pollution prevention project assessments which account for:

- The capacity to account for all cash flows over the project's life; and,
- The ability to integrate the time value of money.

The Net Present Value (NPV), Internal Rate of Return (IRR), and Profitability Indicator (PI) methods meet the above criteria although the NPV is the preferred method because the later two methods may fail to adequately identify the most favorable project. It should be noted that the Payback Period (PP) method does not meet the two criteria listed above and should therefore not be utilized (ASTM, 1992).

Direct Allocation of Costs

It is imperative that industry properly account for the costs associated with pollution in terms of production costs per unit of product. If waste reduction costs are the only production costs considered per unit of product, a bias to waste management (which is generally considered an overhead cost) is created because the waste reduction appears to increase production costs (Zornberg and Wainwright, 1992). Additionally, failure to allocate costs to the products and processes that generate them, results in the inability to identify the following (U.S. EPA, 1992):

- Products and processes responsible for environmental costs
- Financial savings of a prevention program

The accounting system may also be incapable of targeting the opportunity for pollution prevention assessments and investments for high environmental cost products and processes. The proper allocation of production costs is clearly of great importance to the financial analysis. Costs must be assigned to their direct source in a manner which is indicative of how the costs are incurred. Common methods of cost allocation include the use of production inventory data, material balances, or, a combination of the two. ASTM suggests a four-step approach for the performance of the analysis:

- Assemble capital cost data;
- Assemble operational cost data;
- Summarize capital and operational costs, and select financial assumptions; and,
- Perform the profitability analysis.

As mentioned previously, the actual approach taken by industry will be dependent on their specific requirements and the types of analysis to be performed. By examining pollution costs through the use of extended cost inventory and time horizons, long-term financial indicators, and the direct allocation of costs, industry - at a minimum - will gain a better understanding of the actual costs associated with pollution generation. By doing so, industry will target pollution prevention investments and learn how to identify prevention activities which can effectively compete with other company investments. These data will allow for the more accurate estimation of less-tangible costs and benefits in the future.

POLLUTION PREVENTION AT THE DOMESTIC AND OFFICE LEVELS

Concurrent with the Unites States' growth as an international economic superpower during the years following World War II, a new paradigm was established whereby our society became accustomed to the convenience and ease with which goods could be discarded after a relatively short useful life. We have come to expect these everyday comforts with what may be considered an unconscious ignorance towards the ultimate effect of our throw-away lifestyle. In fact many of us, while fearful of environmental degradation, are not aware of the ill-effect our actions have on the world around us. Many individuals who abide by the "Not In My Back Yard" (NIMBY) mind-set also feel pollution prevention does not have to occur "in my house."

The past two decades have seen an increased social awareness of the impact of our lifestyles on the environment. Public environmental concerns include issues such as waste disposal, hazardous material regulations, depletion of natural resources, as well as air, water, and land pollution. Nevertheless, roughly one-half of the total quantity of waste generated each year can be attributed to domestic sources!

More recently, concern about the environment has begun to stimulate "environmentally-correct" behavior among us all. After all, the choices we make today affect the environment of tomorrow. Simple decisions can be made at work and at home which conserve natural resources and lessen the burden placed on our waste management system. By eliminating waste at the source, we are all participating in the protection of the environment by reducing the amount of waste that would otherwise need to be treated or ultimately disposed.

There are numerous areas of environmental concern which can be directly influenced by our actions. The first issue, which is described above, is that of waste generation. Secondly, energy conservation has significantly affected Americans and has resulted in cost-saving measures which have directly reduced pollution. As mentioned previously, energy conservation is directly related to pollution prevention since a reduction in energy use usually corresponds to less energy production, and consequently, less pollution output. A third area of concern is that of accident and emergency planning. Relatively recent accidents like Chernobyl and Bhopal have increased public awareness and helped stimulate regulatory policies concerned with emergency planning. Specifically, Title III of the Superfund Amendments and Reauthorization Act (SARA) of 1986 established the Emergency Planning and Community Right-to-Know Act, and forever changed the concept of environmental management. This law attempts to avert potential emergencies through careful planning and the development of contingency plans. By planning for emergency situations we help protect human health and the environment.

Based on just these three areas of potential concern, a plethora of waste reduction activities can be performed by each of us at home or in an office environment. A few examples follow which are identified by category according to the following notation (Theodore, 1993):

Waste reduction	•
Accidents, health and safety	*
Energy conservation	■

At Home

- • Purchase products with the least amount of packaging
- • Borrow items used infrequently
- * Handle materials to avoid spills (and slips, trips)
- * Keep hazardous materials out-of-reach of children
- ■ Use energy efficient lighting (e.g., florescent)
- ■ Install water flow restriction devices on sink faucets and showerheads

At the Office

- Pass on verbal memos when written correspondence isn't required
- Reuse paper before recycling it
- * Know building evacuation procedures
- * Adhere to company medical policies (e.g., annual physicals)
- ■ Don't waste utilities simply because you are not paying for them
- ■ Take public transportation to the office

The use of "domestic" pollution prevention principles clearly does not involve the use of high-technology equipment or major lifestyle changes; success is only dependent upon active and willing public participation. We can all help to make a difference.

Before pollution prevention becomes a fully accepted way of life, considerable effort still needs to be expended to change the way we look at waste management. A desire to use "green" products and services will be of no avail in a market where these goods are not available. Participation in pollution prevention programs will increase through continued education, community efforts, and lobbying for change. Market incentives can be created and strengthened by tax policies, price preferences, and packaging regulations created at the Federal, State, and local levels. We should all do our part by communicating with industry and expressing our concerns. For example, citizens should feel free to write letters to decision makers at both business organizations and government institutions regarding specific products or legislation. Letters can be positive, demonstrating personal endorsement of a green product, or disapproving, expressing discontent and reluctance to use a particular product because of its negative environmental effect (Theodore, 1993).

By starting now we will learn through experience to better manage our waste while providing a safer and cleaner environment for future generations.

ETHICAL CONSIDERATIONS

Given the evolutionary nature of pollution prevention, it is evident that as technology changes, and continued progress is achieved, our opinion of both what is possible and desirable, will also change. Government officials, scientists, and engineers will face new challenges to fulfill society's needs while concurrently meeting the requirements of changing environmental regulations.

It is now apparent that attention should also be given to ethical considerations and their application to pollution prevention policy. How one makes decisions on the basis of ethical beliefs is clearly a personal issue but one that should be addressed. In order to examine this issue, the meaning of ethics must be known, although the intent here is not to provide a detailed discussion regarding the philosophy of ethics or morality.

Ethics can simply be defined as the analysis of the rightness and wrongness of an act or actions. According to Dr. Andrew Varga, director of the Philosophical Resources Program at Fordham University, in order to discern the morality of an act, it is customary to look at the act on the basis of four separate elements: its object, motive, circumstances, and consequences (Varga, 1979). Rooted in this analysis is the belief that if one part of the act is bad, then the overall act itself cannot be considered good. Of course, there are instances where the good effect outweighs the bad and there may therefore be a reason to permit the evil. The application of this principle is not cut-and-dry and requires us all to make decisions on a case-by-case basis. We must make well-judged decisions rooted in an understanding of the interaction between technology and the environment. After all, the decisions we make today will not only have an impact on our generation, but on many generations to come. If one chooses today not to implement a waste reduction program in order to meet a short-term goal of increased productivity, this might be considered a good decision since it benefits the company and its employees. However, should a major release episode occur which results in the contamination of a local sole-source of drinking water, what is the "good"?

As an additional example, toxicological studies have indicated that test animals exposed to small quantities of toxic chemicals had better health than control groups which were not exposed. A theory has been developed which says that a low exposure to the toxic chemical results in a challenge to the animal to maintain homeostasis; this challenge increases the animal's vigor and correspondingly, its health. However, larger doses seem to cause an inability to adjust resulting in negative health effects. Based on this theory, some individuals believe that absolute pollution reduction might not be necessary.

We must keep an open mind when dealing with these types of issues and facing challenges perhaps not yet imagined. Clearly, there is no simple solution or answer to many of the questions. The EPA is currently attempting to develop a partnership with government, industry, and educators to produce and distribute pollution prevention educational materials. Through continued education and research, we will become better able to address environmental issues while integrating pollution prevention, sound environmental decision making, and the protection of human health and the environment.

Suggested Reading

American Society of Testing and Materials. *Standard Guide for Industrial Source Reduction*. Draft Copy: June 16, 1992.

American Society of Testing and Materials. *Pollution Prevention, Reuse, Recycling and Environmental Efficiency*. June, 1992.

California Department of Health Services. *Economic Implications of Waste Reduction, Recycling, Treatment, and Disposal of Hazardous Wastes: The Fourth Biennial Report*. California: 1988.

Citizen's Clearinghouse for Hazardous Waste. *Reduction of Hazardous Waste: The Only Serious Management Option*. Falls Church: CCHW, 1986.

Congress of the United States. Office of Technology Assessment. *Serious Reduction of Hazardous Waste: For Pollution Prevention and Industrial Efficiency*. Washington, D.C.: GPO, 1986.

Friedlander, S. "Pollution Prevention - Implications for Engineering Design, Research, and Education." *Environment*, May 1989, p.10.

Theodore, L. and Y. McGuinn. *Pollution Prevention*. Van Nostrand Reinhold, New York: 1992.

Theodore, L. and R. Allen. *Pollution Prevention, An ETS Theodore Tutorial*. Roanoke, VA: ETS International, Inc., 1994.

Theodore, L. and Theodore, M. *A Citizen's Guide to Pollution Prevention*. East Williston, NY: 1993.

United States EPA. *Facility Pollution Prevention Guide*. (EPA/600/R-92/088), Washington, DC: May 1992.

United States EPA. "Pollution Prevention Fact Sheets." Washington, D.C.: GPO, 1991.

United States EPA. *1987 National Biennial RCRA Hazardous Waste Report - Executive Summary*. Washington, D.C.: GPO, 1991.

United States EPA. Office of Pollution Prevention. *Report on the U.S. Environmental Protection Agency's Pollution Prevention Program.* Washington, D.C.: GPO, 1991.

World Wildlife Fund. *Getting at the Source.* Executive Summary: 1991.

References

American Society of Testing and Materials, *Standard Guide for Industrial Source Reduction.* Draft Copy: June 16, 1992.

ICF Technology Incorporated. *New York State Waste Reduction Guidance Manual.* (Alexandria, VA: 1989).

Neveril, R.B. "Capital and Operating Costs of Selecetd Air Pollution Control Systems," Gard, Inc., Niles, Ill. (EPA Report 450/5-80-002), December 1978.

Theodore, L. and Theodore, M. *A Citizen's Guide to Pollution Prevention.* (East Williston, NY: 1993).

Theodore, L. Personal Notes, 1993.

Theodore, L. Details available from Louis Theodore, 1994.

Theodore, L. and R. Allen. *Pollution Prevention, An ETS Theodore Tutorial.* (Roanoke, VA: ETS International, Inc., 1993).

Theodore, L. and Y. McGuinn, *Pollution Prevention* (New York: Van Nostrand Reinhold, 1992), p. 171.

Theodore, L. and J.P. Reynolds. *Introduction to Hazardous Waste Incineration.* (New York: Wiley-Interscience, 1989).

Ulrich, G.D. *A Guide to Chemical Engineering Process Design and Economics.* (New York: Wiley-Interscience, 1984).

United States EPA, *The EPA Manual for Waste Minimization Opportunity Assessments.* (Cincinnati, OH: August 1988).

United States EPA, *Pollution Prevention Fact Sheet* (Washington, D.C.: March 1991a).

United States EPA, *1987 National Biennial RCRA Hazardous Waste Report Executive Summary* (Washington, D.C.: GPO, 1991b), p. 10.

United States EPA, *Pollution Prevention Fact Sheet* (Washington, D.C.: March 1991c)

United States EPA, *Facility Pollution Prevention Guide*, (Washington, D.C.: GPO, May 1992), p. 62.

Varga, A. *On Being Human* (Paulist Press, New York: 1978).

Vatavuk, W.M., and R.B. Neveril. "Factors for Estimating Capital and Operating Costs." *Chemical Engineering* (November 3, 1980): pp. 157-162.

Vogel, G.A., and E.J. Martin. "Hazardous Waste Incineration." Part 1. "Equipment Sizes and Integrated-Facility Costs," *Chemical Engineering* (September 5, 1983): pp. 143-146.

Vogel, G.A., and E.J. Martin. "Hazardous Waste Incineration." Part 2. "Estimating Costs of Equipment and Accessories," *Chemical Engineering* (October 17, 1983): pp. 75-78.

Vogel, G.A., and E.J. Martin. "Hazardous Waste Incineration." Part 3. "Estimating Capital Costs of Facility Components," *Chemical Engineering* (November 28, 1983): pp. 87-90.

World Wildlife Fund, *Getting at the Source* (Copyright 1991), p. 7.

Zornberg, R. and B. Wainwright. *Waste Minimization: Applications in Industry and Hazardous Waste Site Remediation* (Riverdale, NY: 1992, © 1993), p. 37.

CHAPTER 2: POLLUTION PREVENTION PROBLEMS

1. Basic Concepts

1.1. *Basic Concepts-1* (unit conversions). [cr] A regulatory agency stipulates that the maximum concentration of benzo(a)pyrene in drinking water should not exceed 200 ng/L. Express this concentration in lb/10,000 U.S. gal.

1.2. *Basic Concepts-2* (basic concepts, ideal gas law, stoichiometry, combustion, volumetric quantities). [cr] Indicate if the following statements are "True" or "False."

1. "S.T.P." refers to the following temperature and pressure conditions: 0°C and 1 atm.
2. The relative humidity is independent of temperature.
3. The stoichiometric equation for the combustion of ethylene is:

$$C_2H_4 + 1.5\ O_2 \rightarrow 2\ CO_2(gas) + H_2O(liquid)$$

4. The U.S. gallon is smaller that the Canadian gallon.
5. The calorimetric bomb gives the "gross heating value" of a combustible material, not the "net heating value."
6. Excess air is defined as "air in excess/total air."
7. For ideal gas mixtures, the volume fraction is equivalent to partial pressure in atm where the total pressure is 1 atm.
8. One gmol of SO_2 weighs 32 g.
9. 100 ppmv = 0.01 vol%.
10. The density of water at 32°F is 62.4 ft^3/lb.

1.3. *Basic Concepts-3* (basic concepts, chemical structure, chemical nomenclature). [dem] Draw the molecular structure and give the systematic name for:

a. Toluene
b. o-Xylene
c. Phenol

1.4. *Basic Concepts-4* (basic concepts, unit conversions). [dem] Express the concentrations for the solutions given below in terms of percentage by weight, ppm, and molarity.

a. 36 g of HCl in 64 cm^3 of water.
b. 0.003 g of ethanol in 1 kg of water.
c. 34 g of ammonia in 2000 g of water.

1.4. *Basic Concepts-4 (continued)*

The following information is needed:

	Densities (kg/L)	Molecular weight (g/gmol)
H_2O	1.0	18
HCl	1.268	36.5
NH_3	0.817	17
C_2H_5OH	0.789	46

% weight = (solute weight /solution weight) x 100
  = [solute weight/ (solute weight+solvent weight)] (100)
ppm = (solute weight, mg) /(solution volume, L)
  = (solute weight, mg)/(solute volume + solvent volume, L)
  = (solute weight, mg)/(solute weight/solute density
    + solvent volume, L)
molarity = solute moles/solution volume
  = solute moles /(solute volume + solvent volume)
  = solute moles /(solute weight/solute density + solvent volume, L)

Note: ppm for liquids is defined on a mass basis, i.e., mg of solute/kg of
     solution.   If the solvent is water, with a specific gravity of 1.0, then
     ppm can also be expressed on a mass/volume basis as indicated
     above.

1.5. *Basic Concepts-5* (basic concepts, Henry's Law constant, solubility,
vapor pressure).   [dem] How much of each of the following gases can be
dissolved in 1 L of water if the gas is in a mixture such that the partial
pressure is 0.26 atm.   Use the following graph to determine the Henry's
Law constant for each gas assuming equilibrium conditions.

  a. Hydrogen at 20°C
  b. Hydrogen at 30°C
  c. Acetylene at 20°C
  d. Acetylene at 30°C

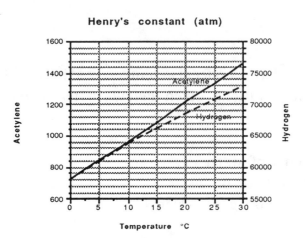

1.6. *Basic Concepts-6* (basic concepts, block diagrams, stoichiometry, steel industry). [jt]   The following process is used to recycle waste from steel pickling liquor.

- Spent pickling liquor exits the steel pickling operation (where scale is removed by reaction to form ferrous sulfate):

$$FeO + H_2SO_4 \rightarrow FeSO_4 + H_2O$$

- Ferrous sulfate is concentrated in an evaporator and fed to a reactor.
- HCl is bubbled through the concentrated liquor to carry out the reaction:

$$FeSO_4 + 2\,HCl \rightarrow FeCl_2 + H_2SO_4$$

- Ferrous chloride is separated from H_2SO_4 in a centrifuge; the H_2SO_4 and unreacted HCl are sent to a degasser.
- The ferrous chloride is roasted to yield useable iron oxide:

$$FeCl_2 + H_2O \rightarrow 2\,HCl + FeO$$

- The HCl from roasting is also sent to the same degasser where recovered H_2SO_4 is recycled to the pickling operation and the HCl is fed to a stripper.
- Inert gases are stripped from the HCl using water, and the HCl is recycled to the ferrous sulfate reactor.

a. Draw a block diagram of the process and label the principal components of all streams.
b. Based on the above stoichiometry (neglecting make-up streams), what is the overall reaction?   What by-products and waste streams are formed?

1.7. *Basic Concepts-7* (basic concepts, thermal pollution, industrial cooling, heat balance). [lar]   Determine the percentage of a river's flow available to an industry for cooling such that the river temperature does not increase more than 10°F.   Fifty percent of the cooling water is lost by evaporation and the return flow is 60°F warmer than the river.

1.8. *Basic Concepts-8* (basic concepts, combustion, coal fired power plant, flue gas). [dks] A coal-fired power plant produces 45 lbmol of exhaust gas per 100 lb of coal burned. Calculate the amount of untreated flue gas (scfm) for a 100-MW power plant. Assume that the heating value of the coal is 7,300 cal/g of coal burned. The overall efficiency of energy conversion is 30%. S.T.P. conditions are given as: T = 60°F and P = 1 atm.

1.9. *Basic Concepts-9* (basic concepts, combustion, coal-fired power plant, sulfur emissions). [dks] Determine the flue gas composition of the uncontrolled emission anticipated from a 100-MW pulverized coal power plant. The following data are available:

Composition of coal:

Component	wt %
Hydrogen	5
Carbon	81
Sulfur	0.64
Nitrogen	1.4
Oxygen	5.76
Ash	6.2

The amount of air used is 15% in excess of stoichiometric requirements. Assume that 95% of all the sulfur in the coal is converted to SO_2 and the balance is converted to SO_3, while 99.5% of all the carbon is converted to CO_2 with the balance being converted to CO. During combustion 20 g of NO_2 are produced per kg of coal burned.

1.10. *Basic Concepts-10* (basic concepts, coal combustion, SO_2 emission, pollution control, Clean Air Act). [ihf] Pollution prevention practices can help a plant to be in compliance with regulations. One of the important regulations is the Clean Air Act Amendments of 1990 (CAA), which limits the emission of sulfur dioxide (SO_2) from coal combustion to 1.2 lb SO_2/MMBtu. An ultimate analysis of a coal provides the percentages by mass of carbon, hydrogen, nitrogen, oxygen, and sulfur in the coal. The higher heating value of a coal (HHV, gross calorific value, gross heating value, or total heating value) may be estimated with reasonable accuracy from the ultimate analysis using the Dulong formula.

$$HHV \ (kJ/kg) = 33{,}801 \ (C) + 144{,}158 \ [(H) - 0.125(O)] + 9413 \ (S)$$

1.10. *Basic Concepts-10 (continued)*

where (C), (H), (O), and (S) are the mass fractions of the corresponding elements. The 0.125 (O) term accounts for the hydrogen bound in the water contained in the coal.

A Pittsburgh bituminous coal with an ultimate analysis of:

C	H	O	N	S	ash (noncombustible)
75.8%	5.1%	8.2%	1.5%	1.6%	7.8%

is burned in a power-plant boiler furnace, and all the sulfur in it (except for any sulfur in the ash) forms sulfur dioxide (SO_2).

a. What is the higher heating value of the coal fuel? Express your answer in MJ/kg and in Btu/lb.
b. Estimate the ratio (mass SO_2 formed)/(higher heating value of the fuel). Express your answer in kg SO_2/GJ, and in lb SO_2/million Btu (or MMBtu). (1 GJ = 1000 MJ = 1,000,000 kJ).
c. Is the plant in compliance with the CAA sulfur emissions? If not what pollution prevention measures would you suggest?

Note: To convert kJ/kg to Btu/lb multiply by 0.4303.
       To convert Kg/GJ to lb/Btu multiply by 2.323

1.11. *Basic Concepts-11* (basic concepts, stoichiometry, neutralization).
[jt]   A semiconductor manufacturing site generates a mixed-acid etch waste stream with the following composition on a wt % basis:

        60 % HNO_3
        20 % HF
        20 % CH_3COOH

Rather than treat this waste, they have agreed to combine it with a waste stream from a nearby acetylene manufacturing site. The acetylene plant uses the carbide process which generates hydrated lime as a by-product according to the following stoichiometry:

$$CaO + 4\,C + 0.5\,O_2 \rightarrow 2\,CO + CaC_2$$

$$CaC_2 + 2\,H_2O \rightarrow Ca(OH)_2 + C_2H_2$$

1.11. *Basic Concepts-11 (continued)*

The two waste streams can be mixed to yield the following saleable products:

Calcium Nitrate - $Ca(NO_3)_2$ - a high quality fertilizer
Calcium Acetate - $Ca(CH_3COO)_2$ - can be used in lieu of lime or gypsum
Calcium Fluoride - CaF_2 - raw material for HF manufacture; flux used in
   steel manufacture

a. Balance the stoichiometric equations for the reaction of hydrated lime with each acid to form the above saleable products.
b. How many lbs of hydrated lime are needed to stoichiometrically treat one lb of the mixed-acid at 100% concentration?

1.12. *Basic Concepts-12* (heavy metals, chromium, treatment, stoichiometry, disposal, conversion factors). [glh] All heavy metals are toxic and must be prevented from entering the environment. Ideally, substitutes can be found for them and source reduction can be achieved. In some instances, recycling can be implemented as a means of pollution prevention. However, if these first two options are not available, then the only recourse is ultimate disposal by landfill. Treatment of heavy metal wastes does not render them non-hazardous but simply makes the metals less available to the environment.

Chromium (VI) is one such heavy metal known to be carcinogenic and must be treated before disposal in a landfill. Often used in oxidation reactions as the chromate ion, CrO_4^{2-}, it also formerly found nearly universal use as a laboratory glassware cleaner. Ideally, the substance is treated in two stages before disposal. First, it is reduced to the Cr^{3+} oxidation state which is less dangerous (Equation 1 below). Then it is precipitated as its insoluble hydroxide using some form of base (Equation 2 below). After filtration of the insoluble $Cr(OH)_3$ salt, it can be disposed of as described in Lunn and Sansone (1990).

a. Balance the equation below and calculate the pounds of sodium metabisulfate required to reduce 5.0 lb of Na_2CrO_4.

$$Na_2CrO_4 + Na_2S_2O_5 \rightarrow Na_2SO_4 + Cr_2(SO_4)_3 \qquad (1)$$

b. Balance the equation below and calculate the kilograms of $Mg(OH)_2$ required to precipitate the chromium(III) as its hydroxide.

1.12. *Basic Concepts-12 (continued)*

$$Cr_2(SO_4)_3 + Mg(OH)_2 \rightarrow Cr(OH)_3 + MgSO_4 \tag{2}$$

c. Assume that the original Na_2CrO_4 is dissolved in 9.3 L of water. Calculate the molarity of the solution.

Lunn, G. and E. B. Sansone (1990), *Destruction of Hazardous Chemicals in the Laboratory*, John Wiley and Sons, Inc., New York.

1.13. *Basic Concepts-13* (conservation of mass; unit conversions; stoichiometry) [smk] It is proposed to use a waste stream, consisting primarily of chlorobenzene, as a fuel for a boiler in a plant. The feed rate of chlorobenzene (C_6H_5Cl) is to be 100 lb/h, and it is to be combusted in 25% excess air. Combustion products are vented out a stack.

a. Write a balanced chemical reaction describing the complete combustion of chlorobenzene in 25% excess air. Use the "typical" composition of air: 20.9% O_2, 79.1% N_2.
b. For every lb of fuel burned, how many lb of air are fed to the boiler?
c. For the feed rates given, what is the mass flow rate (lb/h) of gases leaving the stack?
d. The gases are vented to the atmosphere at a temperature of 400 °F. What is their volumetric flow rate in acfm under these conditions?
e. What is the volumetric flow rate, corrected to standard conditions (1 atm, 25 °C) in units of acfm?
f. What is the volumetric flow rate, corrected to standard DRY conditions in units of dscfm?
g. Convert all inlet and exit mass flow rates to SI units (kg/h).
h. Convert stack gas flows (acfm, scfm, and dscfm) to SI units (m^3/min).

1.14. *Basic Concepts-14* (microbiology, material balances, biodegradation, nutrient requirements, ideal gas law, stoichiometry). [rrd] Biological treatment is an effective and relatively "environmentally friendly" technology for the treatment and control of biodegradable organic waste materials that exist in process waste streams and at uncontrolled hazardous waste sites. This technology can be highly cost effective for the treatment and removal of biodegradable compounds as it takes place at ambient temperature and pressure. It requires much less energy for operation than physical/chemical process options available for

1.14. *Basic Concepts-14 (continued)*

contaminant treatment, i.e., solvent extraction, incineration, solidification, etc. In addition, the end products of biological oxidation of biodegradable chemicals is CO_2, H_2O and other stable materials. This technology requires the intimate contact among all of the reactants (i.e., microorganisms, electron acceptor (oxygen), contaminant, moisture, requisite nutrients, etc.) in order for the process to proceed at a rate and to an extent that is considered effective and efficient from an engineering standpoint. Furthermore, the environment in which biodegradation is taking place must be free of materials inhibitory or toxic to the microorganisms mediating the desired biotransformations.

The requisite mixture of reactants can be estimated based on conducting an oxygen equivalent calculation for the contaminant from stoichiometry, and by making some assumptions regarding the carbon:nitrogen:phosphorous (C:N:P) ratio in cell protoplasm, and energy:synthesis ratios (the fraction of the substrate used by the microorganisms for energy production versus that used as carbon for cell production) for the reaction. A typical oxygen equivalent calculation for hexane is shown below.

First the balanced stoichiometric equation is written assuming O_2 and CO_2 are the end products of the reaction:

$$C_6H_{14} + 9.5\ O_2 \rightarrow 6\ CO_2 + 7\ H_2O \qquad (1)$$

The oxygen equivalent of hexane is defined then as the mass of oxygen required per mass of hexane oxidized:

Hexane oxygen equivalent = [9.5 gmol O_2 (32 g O_2/gmol)]/[1 gmol C_6H_{14} (86 C_6H_{14} g/gmol)] = 304 g O_2/86 g C_6H_{14} = 3.5 g O_2/g C_6H_{14}

The nutrient requirement is generally based on an average C:N:P ratio of 100:20:1 on a weight basis, and it is generally assumed that the net production of cell material is approximately 20% of the substrate removed during biodegradation. The balance of the contaminant is oxidized, generating energy for cell growth and reproduction. When the contaminant is utilized for energy production the oxygen demand described by Equation 1 occurs.

   a. Determine the mass of oxygen required to completely oxidize a 1,000 gal spill of hexane in 1,000 yd^3 of soil. The soil has an available

1.14. *Basic Concepts-14 (continued)*

nutrient pool of 0.03 wt% P and 0.2 wt% N. The specific gravity of the soil is 1.59, while the specific gravity of hexane is 0.659.

b. What is the volume of this oxygen required for complete biodegradation of hexane at STP (20°C, 1 atm) in ft^3? What is the volume of air required in ft^3?

c. Will there be any need to add nutrients to this soil system based on the C:N:P ratio given above, assuming a nutrient application rate of 1.2 times stoichiometric?

1.15. *Basic Concepts-15* (basic concepts, conservation of mass, combustion). [smk] The heat–generating unit in a coal–fired power plant may be simply described as a continuous–flow reactor, into which fuel (mass flow rate F) and air (mass flow rate A) are fed, and from which effluents ("flue gas", mass flow rate E) are discharged.

a. Draw a flow diagram representing this process. Show all flows into and out of the unit. Write a mass balance equation for this process.

b. Suppose the fuel contains a mass fraction y of incombustible component C (for example, ash). Assume that all of the ash is carried out of the reactor with the flue gas (note that in reality, a fraction of the ash generated will remain within the heat-generating unit as bottom ash and must be removed periodically). Write a mass balance equation for component C. What is the mass fraction (z) of C in the exit stream E? What is the effect of increasing the combustion air flow upon z?

c. Suppose a fraction x of the flue gas is recycled to the inlet of the "reactor." (This is commonly done to help suppress formation of pollutants, primarily NO_x). Redraw the flow diagram, including the recycle stream (R). Write a mass balance equation for the overall process, and a mass balance equation around the reactor only. Discuss how these equations are different from the one written in Part a.

d. An air pollution control device is added to the exhaust stream. It is able to remove 99% of the incombustible pollutant C by scrubbing with water. Let S be the mass flow rate of scrubbed material in the water stream. Add this air pollution control unit to the flow diagram, and include all process streams into and out of the unit. Express S in terms of F, A and E.

1.16. *Basic Concepts-16* (economics, reformulation). [wrl] The concept of the time value of money is an integral part of most capital investment decisions, and pollution prevention investments are no exception. The time value (also called *present value*) concept simply states that a dollar today is worth more than a dollar tomorrow. The following problem demonstrates this concept.

An investor may invest $60,000 in either Option A or Option B. The return on the investment for each option is given in the table below. The investor wishes to earn the highest rate of return possible. What is the present value of each option? Assume end-of-year discounting at a rate of 10%.

Table 1. Return on investment for investment Options A and B.

Year	Annual Income Option A	Annual Income Option B
1	$10,000	$10,000
2	$15,000	$10,000
3	$10,000	$15,000
4	$10,000	$15,000
5	$15,000	$10,000
Total	$60,000	$60,000

The present value formula is as follows:

$$PV = AI \frac{1}{(1 + i)^n}$$

where PV = present value of annual income for period n in $, AI = annual income in $, i = annual interest factor or discount rate as a decimal, and n = number of annualized periods

2. Pollution Prevention Principles

2.1. *Pollution Prevention Principles-1* (basic concepts, pollution prevention, definitions). [ihf] Traditionally pollution abatement emphasized pollution control. Today, the shift is towards pollution prevention (P^2). What are the differences between the two terms? When was the concept of P^2 formally introduced by the regulatory agency? (A good source of information is the book by Theodore and McGuinn (1992), *Pollution Prevention*, Van Nostrand Reinhold, New York).

2.2. *Pollution Prevention Principles-2* (regulatory issues, ultimate disposal, liability). [ihf] The Pollution Prevention (P^2) Act of 1990 is the most important regulation regarding pollution prevention. It establishes pollution prevention as a national objective. It sets up a hierarchy of pollution prevention options, which are: Source Reduction, Recycling and/or Reuse, Treatment, and Ultimate Disposal.

Ultimate disposal is the final process in the management of wastes. There are four methods of ultimate disposal. Match the process with the correct definition.

1 - Deep Well Injection
2 - Ocean Disposal
3 - Landfilling
4 - Landfarming

A - The process of disposing hazardous and nonhazardous wastes in the upper layer of the soil. It is an effective and low cost disposal method, and environmentally safe.

B - The process that transfers liquid wastes far underground and away from fresh water sources. It has been predominantly used by the petroleum industry.

C - The process generally used for wastes in the form of sludges. There are two types, area filling and trenching. Both techniques require the use of lime to control odors, and both also produce gas.

D - This ultimate disposal method is probably the simplest of the four techniques but its long term effects are less understood.

2.3. *Pollution Prevention Principles-3* (source reduction, waste minimization, industry) [kg] Explain the following waste reduction techniques that are used in industry:

a. Good housekeeping.
b. Material substitution.
c. Equipment design modification.
d. Recycling.
e. Waste exchange.
f. Detoxification.

2.4. *Pollution Prevention Principles-4* (source reduction, waste minimization, management, waste audits). [kg] The major components of a waste reduction program must include the following: management commitment, communication of the program to the rest of the company, waste audits, cost/benefit analysis, implementation of Pollution Prevention Program, and follow-up. Describe each of these general principles necessary for a successful pollution prevention program.

2.5. *Pollution Prevention Principles-5* (basic concepts, definitions). [smk]

a. Define and discuss the differences between the following terms:
   i. pollution prevention
   ii. pollution control
   iii. waste minimization
b. Consider a degreasing operation in a metal finishing process. This process involves the use of solvents to remove compounds, such as oils, from metal parts to prepare them for further processing. Give an example of modifications that might be made to this part of the process that represent:

   i. source reduction
   ii. recycling
   iii. waste treatment
   iv. waste disposal

2.6. *Pollution Prevention Principles-6* (waste audit, small college, academic labs). [glh] As a rule, point source pollution sites are easily identified (large chemical plants, sewage treatment, power plants, etc.) and can be dealt with more or less satisfactorily by means of regulation

2.6. *Pollution Prevention Principles-6 (continued)*

and/or public pressure. In contrast, discharges of pollution from small quantity generators (homes, restaurants, dry cleaning establishments, etc.) are far more numerous, more widely distributed and therefore more difficult to identify and regulate. It is particularly important that these smaller sources of pollution adopt the desired "Pollution Prevention" attitude.

Small colleges are one example of an intermediate source of pollution in which several departments and offices on campus may use, store and dispose of pollutants quite independently of one another. The four major elements of the Waste Minimization Assessment Procedure are:

1. Planning and Organization
2. Assessment
3. Feasibility Analysis
4. Implementation

Many people overlook the first stage, Planning and Organization, and this oversight limits or even dooms the eventual results of their work. This Planning and Organization stage includes:

1. Gaining management commitment
2. Setting overall goals
3. Organizing the task force.

a. A portion of the Assessment Phase is to select and to prioritize assessment targets. From the list of campus sites below, select the five most likely sources of pollution in this typical small college. First, consider the types and volumes of pollutants likely to be used by the various departments. Then prioritize your choices from 1 to 5 with 1 being most likely to be the largest source of pollution, and 5 being the least likely.

English/History Department	Cafeteria
Library	Chemistry Department
Biology Department	Maintenance/Shop
Dormitories	Psychology Department
Physics Department	Art/Theater Department

Check the suggested solution before continuing. If you disagree, justify your position.

2.6. *Pollution Prevention Principles-6 (continued)*

b. A further part of the assessment phase is to suggest options for dealing with generated wastes. Use your list from above (or the suggested answers) and discuss the various options available for minimizing major waste sources on the small college campus.

U.S.EPA (1990), *Guides to Pollution Prevention--Research and Educational Institutions*, EPA/625/7-90/010, Office of Pollution Prevention, Washington, D.C.

2.7. *Pollution Prevention Principles-7* (Pollution Prevention Act, definitions). [glh] According to the Pollution Prevention Act of 1990, the U.S. policy with regard to pollution prevention can be viewed as the series of options described below:

SOURCE REDUCTION   most preferred

RECYCLING

TREATMENT

ULTIMATE DISPOSAL   least preferred

The EPA has chosen to focus on the first option and only wishes to consider source reduction as a true pollution prevention measure. If this is impossible, then recycling and reuse may be considered.

The definition of Source Reduction is fairly clear. Simply produce less pollution at the source by any combination of strategies available.

Likewise, Recycling (in its many forms) is a well understood principle in today's society, although its true value to our environment seems to be lost on a sizable portion of the population. Any reuse of a substance without chemical change is considered to be recycling. Distillation, melting, filtration and physical changes are permitted.

Treatment can include a number of processes, including physical, biological and chemical methods. Incineration is included in this category as a chemical method of treatment (oxidation).

The misnomer Ultimate Disposal includes deep-well injection, ocean dumping, landfilling, and landfarming.

2.7. *Pollution Prevention Principles-7 (continued)*

a. Consider the list of pollution problems below and briefly discuss which of the options given above would be most applicable. Also discuss the limitations of implementing the other options.

    i. Soil at a railroad station contaminated with PCBs.
    ii. The Greenhouse Effect.
    iii. Volatile organic compounds (VOCs), e.g., xylene, escaping from a stack.
    iv. Ozone layer depletion.
    v. Radioactive wastes from a nuclear power plant.
    vi. Chromium salts from the printing/photographic industry.
    vii. Platinum catalyst in an industrial process.
    viii. Sulfuric acid waste from the pharmaceutical industry.

b. The axiom, "an ounce of prevention is worth a pound of cure" has a long-standing history of applicability to most problems. Discuss its impact in the context of Pollution Prevention with respect to the above-stated EPA policy on the reduction of pollution with its emphasis on source reduction.

2.8. *Pollution Prevention Principles-8* (source classification, emission factors, fugitive emissions). [smk] In order to identify, quantify, and minimize emissions, an inventory must be taken of sources, and estimates or measurements must be made of the magnitude of each source.

Storage and handling emissions occur during the transfer of liquids and solids. One source of such emissions is from storage of petroleum products in tanks. These emissions occur during the filling and emptying of these tanks, and through "breathing losses" due to changes in tank gas volume and liquid volatility as ambient temperature and pressure changes.

Process emissions occur from stacks, vents in reactors, recovery systems, and control equipment.

Fugitive emissions occur from leaking equipment. Leaks may result from improperly sealed as well as damaged equipment, poor maintenance, etc.

Secondary emissions are organic emissions from wastewater collection and treatment systems (such as aeration basins), solvent recovery operations, and accidental spills and leaks.

2.8. *Pollution Prevention Principles-8 (continued)*

Identify the following sources as pollutants, waste, or neither, and classify them as storage and handling, process, fugitive, or secondary emissions. Suggest ways to eliminate or reduce each source.

a. A truck delivers a load of coal to a plant. The coal is dumped into a storage bin at the site, generating coal dust.

b. A solvent bath has a lid that is closed during down–time. When the bath is in use the liquid is exposed to the air.

c. Spent solvent from a bath must be replaced periodically. The used solvent is stored in a drum until it is removed from the site.

d. Solvent is emitted from the stack of a chemical plant. In the atmosphere it undergoes chemical reactions and may contribute to aerosol (smog) formation. Classify both the solvent and the aerosol.

e. In a plating operation, a part is dipped in a chemical bath, and the dipped piece is transferred to a rinse bath several feet away. During the transfer, the dipped piece drips onto the plant floor ("dragout"). Most of the resulting chemical spill evaporates and the rest is washed down the drain, which eventually finds its way into the city sewer system.

f. A pump through which an organic solvent is flowing has a small, undetected leak.

g. A sulfuric acid solution is used to regenerate an ion exchange column.

h. A sulfuric acid stream is neutralized by mixing with a caustic stream from another part of the process. Classify both the reagent streams and the neutralized stream from the process.

Refer to the *Chemical Engineering Progress*, January, 1993, for further details regarding the emission source classification given above.

2.9. *Pollution Prevention Principles-9* (source reduction, waste minimization, solvent vapor recovery, drying ovens). [kg] Solvent vapor emissions from a drying oven are diluted by air to bring the solvent concentrations below the lower explosive level. As a pollution prevention measure you are asked to consider nitrogen as an alternative for air to transport the solvent vapor. What are the advantages of the nitrogen system over air?

2.10. *Pollution Prevention Principles-10* (awareness, attitude, domestic, definitions, resource waste). [glh] According to many experts, Pollution Prevention is an attitude and not a specific discipline or a set of regulations. It is clear that if Pollution Prevention is not a "way of thinking" which is pervasive to one's outlook in everyday activities, then it becomes much more difficult to achieve. Ideally, everyone--from the President of the United States and CEOs of large corporations to blue-collar workers, professionals of every discipline, citizens and even children-- must have a model of everyday life which includes an image of Pollution Prevention whereby the waste and misuse of resources is socially and economically unacceptable.

Industrial pollution can be defined as a deleterious side-product or effect which is generated during a process. Waste of our precious resources is the companion of pollution and should be assiduously avoided. Energy waste is at the head of the list of the misuse of our resources. Ultimately, most pollution can be reduced to energy use terms (how much energy/effort must be expended in order to undo pollution).

As a means of raising an awareness of the pervasiveness of waste within our society and the ease with which significant changes can be made by each of us, complete the following exercise:

a. On the left side of a sheet of paper, make a list of your activities over the last 24 hours (or 48 hours, or one week, etc.) in which people waste energy (electricity, gas, fuel, heat, etc.) and/or other resources that are wasted on a daily basis. Begin with yourself. A list of twenty items should be easily achieved. (Hint: Try not to limit yourself to items in which you had choices but also consider the potential inherent in societal changes, e.g., assume mass transit is available to you. Be creative and think in terms of the best of all possible worlds).

b. On the right side of the same sheet, write a phrase or brief sentence to indicate how you could have (assuming other options were available) been more environmentally conscientious.

Notes to the instructor: This problem can be altered in a number of ways in order to fit your needs. It could be used as a group exercise in class or as a homework assignment. It is important that the student not be inhibited in his/her thinking by such thoughts as "that's too obvious,

2.10. *Pollution Prevention Principles-10 (continued)*

that's too trivial, there was no alternative at the time." The purpose of the assignment is to heighten the student's perception of waste at all levels-from the small to the large. As a follow-up exercise, you might have the students select the five most important incidents of waste on their lists. This ranking of "importance" could be according to the amount of savings that could be realized by changing behavior, along with the potential damange and adverse environmental impact if no change is made.

A third section could be added in which you ask the student (groups) to identify all the forms of pollution (or waste) contained within each entry, especially if it is not obvious, and to identify the reason that it is regarded as pollution. For instance, leaving on unused lights obviously wastes electrical energy. Less obvious perhaps is the fact that heat is generated by the light bulb, adding to the thermal pollution load (air conditioning in the summer). Also, the generation of the electrical power at the power plant may result in carbon dioxide, sulfur and nitrogen oxides, acid rain, mining and refining wastes, radioactive wastes, thermal pollution, to mention a few, depending on the original source of power, whether it be coal, oil, natural gas or nuclear energy.

It is important that students share their various responses to become aware of the power of multiple minds in generating new and creative responses and to impress upon them the plethora of opportunities to prevent pollution.

2.11. *Pollution Prevention Principles-11* (transportation). [dks] Discuss the relative merits of transportation of goods by trucks and by railroad from the point of view of pollution prevention. Be as sepcific as you can in your comments regarding environmental pollution production and energy utilization requirements of each mode of transportation. Identify the type of data needed to provide a quantitative analysis of this problem.

3. Regulations

3.1. *Regulations-1* (laws, regulations). [hb] Discuss the differences between a law and a regulation. Confine your discussion to federal laws and regulations. Interestingly, some regulations are called "rules." There is no legal distinction between a rule and a regulation. Treat them as the same thing in your answer. (Note: many books on law for the layman, law for scientists and engineers, as well as government pamphlets address this topic. One such book is: Coco, A. (1982), *Finding the Law*, Government Institutes, Inc., Rockville, MD.)

3.2. *Regulations-2* (enforcement actions, Pollution Prevention Act). [hb] For the Pollution Prevention Act of 1990:

a. Describe its enforcement provisions.
b. Comment on their effectiveness.

3.3. *Regulations-3* (environmental regulations, goals). [hb] Briefly describe the major thrust or goals or provisions of the following laws: (Note: The texts of the following laws may be found in the United States Code which many libraries possess. Discussions of these laws may be found in the *Environmental Law Handbook*, 1991, or later edition, published by Government Institutes, Inc., Rockville, MD, as well as in many texts on the environment.)

a. The National Environmental Policy Act of 1970 (as amended in 1989) (NEPA); (Public Law 91-190, 42 United States Code section 4321 et seq.)
b. The Resource Conservation and Recovery Act of 1976 (as amended in 1988) (RCRA); (Public Law 94-580, 42 United States Code section 6901 et seq.)
c. The Comprehensive Environmental Response, Compensation and Liabilities Act of 1980 (CERCLA or Superfund); (Public Law 96-510, 42 United States Code section 9601 et seq.)
d. The Superfund Amendments and Reauthorization Act of 1986 (SARA) Title III - also known as The Emergency Planning and Community Right-to-Know Act; (Public Law 99-499, 42 United States Code section 11001 et seq.)
e. The Pollution Prevention Act of 1990 (Public Law 101-508).

3.4. *Regulations-4* (hazardous wastes, regulatory definitions, waste classification). [ihf] The code of Federal Regulation, Title 40 establishes a classification of hazardous waste. (ihf)

"D" wastes, or characteristic wastes, are defined by 40 CFR Part 261, Subpart C. These are classified on the basis of the following characteristics; ignitability, toxicity, reactivity, and corrosivity. If the waste is classified by its ignitability it carries the "D001" code. Some common D codes are:

D code	Classification basis
D001	Ignitability
D002	Corrosivity
D003	Reactivity
D004	Toxicity characteristic leaching procedure (TCLP)

Here are some waste streams. Identify the waste classification for each:

a. A concentrated sodium hydroxide solution.
b. A waste stream that is packed in light drums. If exposed to air it will undergo an exothermic reaction and catch fire.
c. A waste containing high concentrations of leachable polychlorinated biphenyls (PCBs).

3.5. *Regulations-5* (regulatory issues, small quantity generators). [ihf] A 1985 nationwide survey of small quantity hazardous waste generators defined Small Quantity Generators (SQGs) and Very Small Quantity Generators (VSQGs). Give the definitions of SQGs and VSQGs. Discuss the challenges facing SQGs. (Reference: Theodore, L. and Y. C. McGuinn, Pollution Prevention, Van Nostand Reinhold, NY. 1992).

3.6. *Regulations-6* (Resource Conservation and Recovery Act (RCRA), mixture rule, "Derived From" rule). [lar] Under RCRA there are two key provisions which were adopted without adequate public input. The U.S. Circuit Court of Appeals for the District of Columbia has ordered EPA to reconsider these provisions. This has resulted from the case of Shell Oil Company versus U.S. EPA. The two provisions are: the "mixture" rule; and

3.6. *Regulations-6 (continued)*

the "derived from" rule. Discuss these rules and how pollution prevention relates to these provisions of the regulations. (References: *The National Environmental Journal*, March/April 1992; *Chemical Engineering Progress*, May 1992).

3.7. *Regulations-7* (fugitive air emission, Clean Air Act, VOCs). [dlu] Summarize the regulations and standards of performance for fugitive emission sources of volatile organic compounds (VOCs) in petroleum refineries. Summarize the sources of fugitive emissions and the suggested or required remediation scheme(s). Give an example or two to illustrate the principles.

References:
1. *Federal Register* Vol. 48, No. 2, p 279 (Tuesday, Jan. 4, 1983).
2. U.S. EPA (1984), *Air Pollution Training Institute, Course 482, Sources and Control of Volatile Organic Air Pollutants*, EPA 480/2–84–001. p. 197.
3. 40 CFR Part 60, Subpart G.
4. U.S. EPA (1991), *Control Technologies for Hazardous Air Pollutants*, EPA/625/6-91/014. pp. 3-7.

3.8. *Regulations-8* (regulatory issues, compliance, delisting, "Derived From" rule). [wrl] You are Director of Environmental Affairs for Biguns Chemical Co. One of your engineers rushes into your office to tell you that he has the solution to all of Biguns' hazardous waste disposal problems. This engineer has just returned from lunch with an equipment vendor who has a proprietary process proven to render non-hazardous virtually any EPA-listed hazardous waste. Prior to firing the employee (or at least before administering the breathalizer test), please explain why this is impossible.

3.9. *Regulations-9* (regulations, Clean Air Act). [hb] In 1990, Congress amended the Clean Air Act. These amendments have in effect created and entirely new law. The previous three Titles have been expanded into 11 Titles. Many of the pre-1990 provisions were changed, expanded and/or strengthened. In addition, many completely new provisions and programs have been inserted into the law. The 1990 Clean

3.9. *Regulations-9* *(continued)*

Air Act Amendments have produced a law that is unusually lengthy, detailed, varied and complex. It has the potential to significantly affect the construction of new facilities and to require the modification of existing facilities.

Briefly summarize the issues addressed by, and the major provisions of, the first seven Titles of the Clean Air Act Amendments of 1990. (Hint: your answer should include reference to National Ambient Air Quality Standards (NAAQSs), attainment, nonattainment, reasonably available control measures (RACM), maximum achievable control technology (MACT), major sources, area sources, toxic air pollutants, acid rain, allowances, control technique guidelines (CTGs), permits, stationary sources, mobile sources, chlorofluorocarbons (CFCs), and enforcement).

3.10. *Regulations-10* (criminal penalties, Department of Justice). [hb]

Criminal prosecutions for violations of environmental laws are conducted by the Department of Justice. List as many factors you think the Department of Justice may consider when deciding whether or not to conduct a <u>criminal</u> prosecution against a violator. (Hint: Think of factors that show how irresponsible or negligent the violator was rather than what laws or regulations were violated).

3.11. *Regulations-11* (regulatory issues, definitions, liability). [ihf]

The term liability is very often used with environmental regulations. Related terms are strict liability, joint and several liability, retroactive liability, and cradle-to-grave liability. In addition, other terms often used in connection with environmental regulations and enforcement include negligence, trespass and nuisance. Briefly explain these terms. What are their implications in pollution prevention? References: 1) *Black's Law Dictionary*, 2) *Environmental Law Handbook* (1991), Government Institutes, Inc., Rockville, MD.

3.12. *Regulations-12* (regulatory issues, definitions, groundwater) [ap] Determine whether the statements below are true or false. The following abbreviations are used:

CERCLA - Comprehensive Environmental Response, Compensation, and Liability Act
CWA - Clean Water Act
EWRA - Emergency Wetlands Resources Act
RCRA - Resource Conservation and Recovery Act
SARA - Superfund Amendments and Reauthorization Act
SDWA - Safe Drinking Water Act
TSCA - Toxic Substances Control Act
USEPA - United States Environmental Protection Agency

1 - The SDWA regulates water delivered by public water suppliers, not the protection of drinking water supplies.
2 - Both SDWA and RCRA authorize EPA to bring injunctive actions against potential violators to protect groundwater from contamination in cases of imminent and substantial endangerment of public health.
3 - The wellhead protection program to protect wells serving public water systems was implemented in the CWA.
4 - The "dredge and fill" material permit program for wetland protection administered by the US Army Corps of Engineers is contained within the EWRA
5 - The use and disposal of sewage sludge is regulated by CERCLA.
6 - The "cradle to grave" concept in RCRA requires the creation of documents tacking the generation and flow of solid wastes signed by "the generator and the disposal facility" only.
7 - A major focus on groundwater protection and cleanup has been provided by RCRA.
8 - The chemical production industry has endorsed the front end regulatory philosophy of TSCA, which requires pre-market evaluation of chemicals. As a result the concept has been a remarkable legislative success.
9 - According to CERCLA, application of fertilizers for farming does not constitute a release of hazardous waste to the environment.
1 0- The two federal statutes specifically dealing with underground storage tanks are RCRA and SARA.

3.13. *Regulations-13* (regulatory issues, compliance, liability)   [wrl]
You are Director of Environmental Affairs for Biguns Chemical Co. Your environmental compliance auditor has just informed you that one of your engineers has, at his own initiative, been using a small (1,000 barrels) spare storage tank to collect various low pH (pH of 1 to 2) wastes from around the plant. The engineer has kept careful records, has been careful not to mix incompatible compounds and has not collected any listed hazardous wastes. Once a month, the engineer tests the contents of the tank and then adjusts the pH to make it neutral, at which time the waste is properly disposed.

You immediately fire the engineer and inform the appropriate state and federal environmental agencies of the incident.   Why?

3.14. *Regulations-14* (regulatory issues, compliance, hazardous waste definition, recycling). [wrl]   A confidential survey of the chemical manufacturing industry was conducted in Louisiana in an attempt to identify possible impediments to hazardous waste recycling unique to the state. One problem identified related to the classification of sludges and by-products exhibiting hazardous waste characteristics. Under RCRA, these by-products and sludges are not considered solid waste when reclaimed. Table 1 below is an excerpt from 40 C.F.R. § 261.2. Items indicated with a <u>Yes</u> are solid waste. Any item marked <u>No</u> would not be solid waste when reclaimed or accumulated and, therefore, not subject to the stringent requirements of EPA's Treatment, Storage and Disposal (TSD) rules for recycling. The nature of these rules makes compliance expensive.

Table 1. Federal solid waste designation form 40 CFR 261.2.

	Use constituting disposal	Burning for energy recovery or use to produce a fuel	Reclamation	Speculative accumulation
Spent materials (both listed and nonlisted/characteristic)	Yes	Yes	Yes	Yes
Sludges (listed)	Yes	Yes	Yes	Yes
Sludges (nonlisted/characteristic)	Yes	Yes	No	Yes
By-products (listed)	Yes	Yes	Yes	Yes
By-products (nonlisted/characteristic)	Yes	Yes	No	Yes
Commercial chemical products listed in 40 CFR § 261.33 that are not ordinarily applied to the land or burned as fuels.	Yes	Yes	No	No
Scrap metal	Yes	Yes	Yes	Yes

Yes - Defined as a solid waste
No - Not defined as a solid waste

3.14. *Regulations-14 (continued)*

Louisiana chose, as many states did, to adopt the RCRA rules verbatim with a few minor exceptions, for example, Table 1. Table 2 below is an excerpt from the Louisiana Administrative Code (LAC 33:V.109) and is nearly identical to Table 1 with just two exceptions relating to sludges and by-products exhibiting characteristics of hazardous waste and intended for reclamation (items marked with * are wastes).

Table 2. Louisiana's solid waste designation from LAC 33:V.109.

	Use constituting disposal (1)	Burning for energy recovery or use to produce a fuel (2)	Reclamation (3)	Speculative accumulation (4)
Spent materials	*	*	*	*
Sludges (listed in LAC 33:V.4901)	*	*	*	*
Sludges exhibiting a characteristic of hazardous waste	*	*	*	*
By-products (listed in LAC 33:V.4901)	*	*	*	*
By-products a characteristic of hazardous waste	*	*	*	*
Commercial chemical products listed in LAC 33:V.4901.E and F	*	*		
Scrap metal	*	*	*	*

* - Defined as a solid waste

From Table 2, it can be seen that recycling of sludges and by-products exhibiting characteristics of hazardous waste would be regulated as hazardous waste and not as recyclable products.

The survey identified two companies that produced more than 6,000,000 pounds of characteristic by-products which could be recycled but were not. The respondents acknowledged that the same by-products were being produced and economically recycled at the companies' plants in other states where no TSD license was required. No recycler was interested in establishing a TSD licensed facility to handle the material since it would have to compete against recyclers not required to bear the expense of TSD compliance. Since the material is legally declared hazardous in Louisiana, it falls under manifest requirements and can not be shipped to non-licensed recycling facilities, even those out of state. The result of the regulation is 6,000,000 pounds of waste material going to hazardous waste disposal, either landfill or deep-well injection, in Louisiana while the same material is being successfully recycled in Texas and Ohio. Industry estimates the annual loss on this by-product (loss of

3.14. *Regulations-14* *(continued)*

material plus cost of disposal) to be in the range of $1,000,000 to $2,000,000 per year.

If you were the environmental manager for one of the companies affected by this regulation, what options might you have in dealing with the problem?

3.15. *Regulations - 15* (maritime pollution prevention, MARPOL). [hb] The United States is a Party to the International Convention (i.e. treaty) for the Prevention of Pollution From Ships. The Convention was adopted by the International Maritime Organization (IMO) in 1973. The Convention was modified by the "Protocol" of 1978. In short, the Convention is known as MARPOL 73/78. The Convention is divided into five parts called Annexes. Give a brief description of the pollution concerns addressed by each of these five Annexes.

NOTE: The updated text of the Convention was published in 1992 by the IMO, 4 Albert Embankment, London SE1 7SR under the title MARPOL 73/78 (publication no. IMO-520E). MARPOL 73/78 is also available from Labelmaster, 5724 N. Pulaski Road, Chicago, IL 60646-6797. Phone no. 800-621-5808. Cat.no. MARPOL-91, $110. MARPOL 73/78 may also be available through interlibrary loan from the U.S. Merchant Marine Academy, Kings Point, NY, SUNY Maritime College, Bronx, NY, the State Maritme Academies in California, Maine, Massachusetts, and Texas as well as other libraries participating in the national OCLC network. Discussions of the provisions of MARPOL 73/78 can be found in many textbooks on the topics of: the environment, the oceans, pollution, and shipping. The following are some examples of books where discussions of MARPOL 73/78 can be found:

Cormack, D. (1983), *Response to Oil and Chemical Marine Pollution*, Elsevier Science Publishing Co., New York.
Gerlach, S.A. (1981), *Marine Pollution: Diagnosis and Therapy*, Springer Verlag, New York.
Lidgren, K. and Norrby, S. (1980), *Oil and Chemical Pollution From Ships: Economics and Control Measures*, Ministry of Agriculture, S-103 33 Stockholm, Sweden.
Timagenis, Gr. J. (1980), *International Control of Marine Pollution* (2 vol), Oceana Publications Inc., Dobbs Ferry, New York.

3.16. *Regulations - 16* (maritime pollution prevention, MARPOL, oil pollution, ocean pollution). [hb]   The International Convention For The Prevention of Pollution From Ships, to which the United States is a Party (see NOTE in Problem 3.15 above), specifies the conditions under which an oil tanker may discharge oil into the sea.   List these conditions.

3.17. *Regulations - 17* (regulatory issues, emission, mass balance, automobile coating). [ihf]   In a plant, the coating of an automobile is carried out in two steps.   A primary coat is first applied, followed by a top coat.   Regulations indicate that the total allowable VOC emission from the operation is 80 kg/h.   The top coat contains 520 kg VOC/m^3 of coating material-water (CM-water), while the prime coat contains 300 kg VOC/m^3 CM-water.   Assume that the reasonably available control technology (RACT) transfer efficiency is 100% in the coating, and all the VOC evaporates.   The plant uses 6 L of prime coat CM-water/car and 4 L of top coat CM-water/car for the top coat. The plant is to coat one car/min.

The bubble concept was introduced with the Clean Air Act (CAA) Amendments of 1977.   This concept allowed industries to treat emissions from a plant (or operation) by superimposing a "bubble" around the plant for which there might be a host of individual emissions.   By enforcing the regulation based on total emissions, industry was provided the option of controlling individual discharges from various locations and processes in the plant in the most convenient and/or economic manner to yield a cumulative emission rate specified by permit.

Determine if the above coating operation has a viable "bubble," i.e., are the two individual emissions at or below the total allowable emission rate of 80 kg/h?

4. Source Reduction

4.1. *Source Reduction-1* (VOC, spray transfer efficiency, coatings). [smk] A surface coating facility must reduce VOC emissions in order to comply with new state regulations. Options for VOC emissions reduction include:

1. reformulation of the coating
2. process modification
3. implementation of emissions controls

Using the data that follow, evaluate the percentage reduction in VOC emissions that the company can realize by implementing the relatively simple options of:

a. (without process modification) only reformulation of coating
b. (without reformulation) only process modification from spray to electrostatic spray
c. both reformulation and process modification

Assume that coating thickness and throughput are unchanged.

Data:
Current coating formulation: 25 vol% solids, 5.5 lb VOC /gal coating
Proposed reformulation: 35 vol% solids
Solvent density (unchanged in reformulation): 7.36 lb VOC/gal VOC
Transfer efficiency for current spray process: 0.45
Transfer efficiency for revised process: 0.75
Current coating usage: 10 gal/h

4.2. *Source Reduction-2* (electroplating, drag-out, source reduction). [jt] A nickel electroplating line uses a dip-rinse tank to remove excess plating metal from a specific part line. Currently, a single tank is used which requires R gal/hr of fresh rinsewater to clean F parts/hr (see figure below). Each part contains f ounces of metal residue. The residue in the tank is expressed as r ounces of metal residue/gal of bath. Assume the cleaning in stage i is governed by the equilibrium relation:

$$\lambda = \frac{f_i}{r_i} = \frac{(\text{ounces of metal residue/part})}{(\text{ounces of metal residue/gal bath})}$$

4.2. *Source Reduction-2 (continued)*

   a. Calculate the reduction in rinsewater flowrate as compared to the single-stage unit if a two-stage countercurrent rinse tank is used, and 99% of the residue must be removed, i.e., $(f_{in} - f_{out})/f_{in} = 0.99$. Assume the drag-out volume is negligible, i.e., parts leave the bath practically dry.

   b. Has the total metal content of the exit rinsewater been altered? Discuss implications for further wastewater treatment/reuse.

One stage:

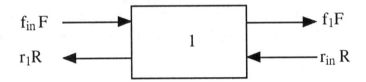

Two stage:

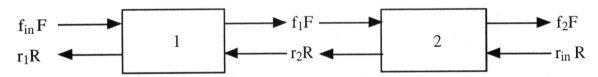

4.3. *Source Reduction-3* (stockroom, chemicals, strategies, source reduction). [glh] The reduction of waste chemicals is perhaps the most important component of pollution prevention. Often in the academic and industrial laboratories, unused chemicals must be disposed of and, aside from the polluting aspects of any form of disposal, this can be very expensive. It may cost the chemistry department of a small college nearly $20,000 for a one-time disposal of the chemicals that have accumulated over decades. Industrial research labs routinely have a certified disposal agent come into the workplace on a monthly basis to pack up and remove hazardous wastes. The cost will vary greatly depending on the nature of the industry, but an annual cost of $40,000 is a reasonable figure for a small research lab in a chemical company.

Typical costs for disposal of hazardous wastes are:

Lab-Pack - a 55 gal drum in which bottles and cans are packed in an inert adsorbent bed for transport will hold approximately 12 one-gal bottles. Cost for disposal ≈ $350.

4.3. *Source Reduction-3 (continued)*

55 gal drum of halogenated solvents.   Cost for disposal ≈ $475.
55 gal drum of flammable, non-halogenated solvents.   Cost for disposal ≈ $265.
55 gal drum of corrosives (acid, base, etc.).   Cost for disposal ≈ $325.

It must be kept in mind that disposal costs will continue to escalate in the foreseeable future.   This is a prime motivator for large and small waste generators alike.

There are a variety of strategies to minimize waste, cost and pollution. Below are five sources of wastes which require disposal in industrial and academic settings.   Discuss ways in which these may be minimized.

1. A 500 g bottle of a hazardous organic compound is 5 years old, still contains 425 g but is no longer needed.
2. Three 100 g bottles of an unknown substance are found alphabetically in the "P-Q-R" section of the stockroom.   The labels have disintegrated.
3. Trichloroethylene, a known carcinogenic hazard, is a liquid, and is used extensively as a solvent.   Several 5 gal cans are kept on hand. Much or most of what is used probably evaporates.
4. LPZ Industries, Inc., a local corporate friend of the community, generously donates its excess unused chemicals to local schools and colleges.
5. Corrosive materials, such as sodium hydroxide solution and sulfuric acid solution are routinely washed down the sink.

4.4. *Source Reduction-4* (Chromium, substitution, oxidation, stoichiometry, equation balancing, source reduction, conversion factors, density) [glh]   Chromium wastes have a long history of simply being "dumped" into the nearest sewer where they inevitably end up as a toxic pollutant of our environment.   With the resulting awareness generated during the past few decades of environmental concern, minimizing these wastes by source reduction has become essential.   There is no way to destroy chromium waste products and they must eventually find their way into a landfill.

4.4. *Source Reduction-4 (continued)*

One such method of source reduction is for manufacturers to find a substitute for chromium(VI) which will accomplish the same purpose. When its use is as an oxidizing agent, there are several choices available to the chemists. Common household bleach, NaOCl, is often an acceptable substitute for other oxidizing agents and generally generates no serious pollutant.

   a. Cyclohexanone, $C_6H_{10}O$, is an intermediate in the production of nylon and is also used as an industrial solvent. It can be synthesized from cyclohexanol, $C_6H_{12}O$, by oxidation with an acidic aqueous solution of sodium chromate, Na_2CrO_4. Balance the equation below and calculate the number of lb of sodium chromate needed to produce 1000 lb of cyclohexanone per day. Assume the reaction is 100% complete.

$$Na_2Cr_2O_4 + C_6H_{12}O \rightarrow C_6H_{10}O + NaH_2CrO_3 + H_2O$$

   b. After treatment of the waste chromium with alkali to produce an insoluble precipitate, $Cr(OH)_3$, the substance must be disposed of in a landfill. Calculate the volume in ft^3 of chromium hydroxide sent to the landfill by this process each year. Assume that the density of the precipitate is approximately 2.9 g/cm^3 and that the plant produces cyclohexanone 200 d of the year.

   c. Balance the equation below and calculate the volume (in gal) of bleach required per year to produce 1000 lb of cyclohexanone per day. Household bleach is 5.25% NaOCl by weight and 1 gal of the solution weighs 8.0 lb.

$$C_6H_{12}O + NaOCl \rightarrow C_6H_{10}O + NaCl + H_2O$$

4.5. *Source Reduction-5* (material balances, solvent extraction, academic laboratories). [rrd] In many environmental analysis laboratories, methylene chloride is utilized for the extraction of semi-volatile organics (SVOs) from liquid and solid environmental samples. Typically in supercritical fluid extraction, SVOs in a 50 g soil sample or a 100 mL liquid sample are extracted with 250 mL of methylene chloride in sequential soxhlet or liquid/liquid extraction steps. The extracts are then dried over sodium sulfate and condensed in a Kaderna-Danish concentrator to a 5 mL final sample volume. The procedure is tedious, the glassware is expensive, the solvent is hazardous, and disposal of the samples after analysis is expensive.

4.5. *Source Reduction-5 (continued)*

One alternative under investigation for use in environmental analysis is supercritical fluid extraction (SFE) with supercritical CO_2. This extraction procedure involves high pressure and moderate temperature CO_2 for the extraction of SVOs from 2 g soil samples, and their transfer into small volumes (e.g., 10 mL original volume producing a final sample volume of 8 mL) of cooled methanol. Not only does this provide economy in analytical costs and effort, it provides significant advantages to the laboratory in terms of waste reduction and improvements in worker safety.

a. Perform a material balance for the soxhlet and supercritical fluid extraction procedures using a 1,000 sample/yr basis.
b. Comment on the relative impact of the effluent stream not "recovered" in the concentrated sample used for analysis in terms of potential environmental impacts and (perhaps) eventual pollution control costs.
c. For the soxhlet process calculate the yearly methylene chloride evaporated, while for the SFE process, calculate the yearly emission of methanol.

4.6. *Source Reduction-6* (material balance, cost effectiveness, institutional barriers). [rrd] Based on Problem 4.5, answer the following questions:

a. Evaluate the cost effectiveness of the switch to SFE for a moderate sized environmental laboratory that processes 1,000 solid samples/yr if the capital and operating costs for each option are as shown below.

Operating data for both the soxhlet and SFE extraction processes are summarized below:

Soxhlet Extraction
    Glassware costs = $1,000/unit
    Average unit life = 2 years
    Methylene chloride costs = $10/L
    Sample disposal costs = $50/gal
    Analysis rate = 8 samples/8 hr d

4.6. *Source Reduction-6 (continued)*

SFE Unit
  Equipment costs = $30,000 for an eight sample unit
  Average unit life = 7 years
  SFE CO_2 = $250/cylinder; 150 samples/cylinder
  Sample disposal costs = $10/gal
  Expendables cost = $2/sample
  Analysis time = 32 samples/8 hr d

Technical labor = $25/hr for an 8 hr working day

b. Why might there be institutional resistance to this extraction method shift?

4.7. *Source Reduction-7* (emission factors, fugitive emissions, fuel storage tanks). [rrd] The U.S. Air Force uses thousands of above ground storage tanks for the intermediate storage of Jet naphtha (JP-4) prior to filling their planes on the flightline. Most are currently covered with a light grey Air Force issue paint on their roofs and shells.   There is some concern that these tanks may be significant VOC emitters, and that costly control systems will have to be implemented to achieve even moderate levels of control for these fixed roof tanks. You have been tasked with:

1. Making preliminary estimates of the levels of annual release from these tanks on a lb/yr and % throughput basis, and
2. To evaluate potential control techniques that can be implemented to minimize tank emissions by at least 10%.

You quickly look for an EPA reference on estimating emission releases from Treatment Storage and Disposal Facilities and come across the document EPA 560/4-88-002, December 1987, "Estimating Releases and Waste Treatment Efficiencies for the Toxic Chemical Release Inventory Form," Office of Pesticides and Toxic Substances, Washington, D.C. In this document you find the following equations that predict breathing losses (vapor release through vapor expansion) and working losses (vapor release during tank filling operations):

4.7. *Source Reduction-7 (continued)*

Breathing Loss:

$$L_B = 2.26 \times 10^{-2} \, (M_v) \, [P/(P_a - P)]^{0.68} \, (D)^{1.73} \, (H)^{0.51} \, (\Delta T)^{0.50} \, (F_p) \, (C) \, (K_c)$$

where
M_v = molecular weight of vapor stored in the tank (lb/lbmol)

P = true vapor pressure at bulk liquid conditions (psia), see Tables 1 and 2 below

P_a = average atmospheric pressure at the tank location (psia)

D = tank diameter (ft)

H = average vapor height including roof volume; if no data are available regarding conical roof heights, assume H = 1/2 D (ft)

ΔT = average ambient diurnal temperature change = Average maximum - Average minimum temperature (°F)

F_p = paint factor (dimensionless) see Table 3 below

C = adjustment for small diameter tanks (dimensionless) = 0.54 @ 10 ft, 0.75 @ 15 ft, 0.90 @ 20 ft, 0.96 @ 25 ft, 1.0 @ 30 ft

K_c = product factor (dimensionless) = 0.65 for crude oil, 1.0 for all other liquids

Table 1. Physical Properties of Jet Naphtha, JP-4. (EPA 560/4-88-002)

M_v	Product Density @ 60°F (lb/gal)	Vapor Density @ 60°F (lb/gal)	Vapor Pressure (psia) at Average Storage Temperature, Ts				
			40°F	50°F	60°F	70°F	90°F
80	6.4	5.4	0.8	1.0	1.3	1.6	2.4

Table 2. Average Storage Temperature as a Function of Tank Paint Color. (EPA 560/4-88-002)

Tank Color	Average Storage Temperature, Ts
White	$T_A + 0$
Aluminum	$T_A + 2.5$
Gray	$T_A + 3.5$
Black	$T_A + 5.0$

4.7. *Source Reduction-7 (continued)*

Table 3. Paint Factors for Fixed Roofed Tanks. (EPA 560/4-88-002)

Tank Color		Paint Factors (F_p) Paint Condition	
Roof	Shell	Good	Poor
White	White	1.00	1.15
Aluminum(specular)	White	1.04	1.18
White	Aluminum(specular)	1.16	1.24
Aluminum(specular)	Aluminum(specular)	1.20	1.29
White	Aluminum(diffuse)	1.30	1.38
Aluminum(diffuse)	Aluminum(diffuse)	1.39	1.46
White	Gray	1.30	1.38
Light Gray	Light Gray	1.33	1.44
Medium Gray	Medium Gray	1.40	1.58

Working Loss:

$$L_w = 2.40 \times 10^{-5} \, (M_v) \, (P) \, (V) \, (N)(K_n)(K_c)$$

where V = tank capacity (gal)

N = number of turnovers/yr = (throughput, gal/yr)/V

K_n = turnover factor (dimensionless); $N < 36$, $K_n = 1.0$;
$N = 50$, $K_n = 0.7$; $N = 100$, $K_n = 0.47$; $N = 150$, $K_n = 0.35$;
$N = 250$, $K_n = 0.28$; $N = 350$, $K_n = 0.25$.

The tanks you are concerned with have the following general characteristics:

Tank diameter = D = 20 ft   Tank height = H = 32 ft
Nominal tank volume = V = 50,000 gal
Tank color = light grey   Fuel = Jet naphtha (JP-4)
Daily throughput = 12,500 gal/d
Average ambient T = 40 °F   Average maximum T = 60 °F
Average minimum T = 30 °F
Atmospheric pressure = 13.5 psi

With these input data answer the two questions posed to you. In addition, suggest the implementation of a specific control technique to meet the 10% reduction goal.

4.8. *Source Reduction-8* (emission factors, fugitive emissions, fuel storage tanks). [smk] Substantial amounts of volatile organic chemicals (VOCs) may be lost to the atmosphere in storing and handling them using storage tanks.

a. Estimate the total loss of gasoline from a storage tank (80 ft diameter and 40 ft height) with a fixed roof configuration. The total loss consists of breathing loss (L_B) and working loss (L_W). For a fixed roof tank, these losses may be estimated from the equations given in Problem 4.7.

The following numerical values are given for the system under consideration:

K_n = turnover factor = 1.0; P = true vapor pressure = 7 psia†
total throughput/yr = 10^6 bbl/yr
average ambient temperature = 65°F
average maximum daily temperature = 70°F
average minimum daily temperature = 50°F

b. The amount of VOC lost can be significantly reduced by installing a freely vented floating roof tank. Estimate the total loss L_T (lb/yr) from a internal floating roof tank using the following relationship:

$$L_T = L_R + L_W + L_F + L_D$$

where, L_R = loss from rim seal, lb/yr; L_W = loss from withdrawal, lb/yr; L_F = deck fitting loss, lb/yr; L_D = deck seam loss, lb/yr

L_R, L_W, L_F, and L_D may be estimated from the following equations:

$$L_R = K_S v^n P^* D M_v K_C$$

where, K_S = seal factor, 2.5; v = average air velocity in mi/hr = 10 mi/hr; n = seal related wind speed exponent = 0; P^* = vapor pressure function = $P/P_A/\{1+(1 - P/P_A)^{0.5}\}^2$; P_A = atmospheric pressure; L_W = 0.943 (Q C W_L/D) $\{1 + (N_C F_C/D)\}$; Q = throughput in bbl/yr; C = shell clinging factor = 0.0015; W_L = average organic liquid density in lb/gal = 5.6 lb/gal; N_C = number of support columns = 1; F_C = effective column diameter in ft = 1 ft; M_v = molecular weight of vapor = 62; K_C = product factor = 0.65 for crude oil and 1.0 for all other liquids; and

4.8. *Source Reduction-8 (continued)*

$$L_F = F_F \, P^* \, M_v \, K_C$$

where F_F = fitting factor in lbmol/yr = 300 lbmol/yr

and

$$L_D = K_D \, S_D \, D^2 \, P^* \, M_v \, K_C$$

where K_D = deck seam loss per unit seam length factor (lbmol/ft–yr)
= 0 lbmol/ft–yr; S_D = deck seam length factor in ft/ft^2.

† True vapor pressure is obtained from the stored liquid temperature
which depends on the ambient temperature as well as the type of
paint on the storage vessel. Here, fresh white paint on the surface
has been assumed so that the storage temperature is the same as the
ambient temperature (see Table 2, Problem 4.7).

4.9. *Source Reduction-9* (source reduction, waste minimization,
economics, net present worth, internal rate of return). [ihf] A key element
in pollution prevention is source reduction to minimize waste generation.
This may necessitate changes in the production process. To justify the
economic soundness of the changes, a net present worth (NPW) and a
discounted cash flow (DCF) internal rate of return are performed. NPW
should not be negative. If a process change requires an initial total
investment at year 0, I_0, and will generate an annual cash flow (in this case
annual savings at year j), CF_j, over a period of N years (the economic life of
the plant), then the NPW is calculated as:

$$NPW = -I_0 + \sum_{j=1}^{N} \frac{CF_j}{(1 + r)^j}$$

where r = interest rate expressed as a fraction.

The DCF internal rate of return is defined as the rate of return that
results in a zero NPW. Substitution of NPW = 0 in the above equation
yields an equation with one unknown, r. The DCF internal rate, r, is then
obtained by trial-and-error.

4.9. *Source Reduction-9* (continued)

The MIF facility is considering a process modification to reduce the wastewater generated. This modification requires a total investment of $8,000,000. The anticipated saving in waste treatment costs is $3,000,000/yr. The economic life of the process is 4 yr. Calculate:

a. The net present worth (NPW) if the interest rate is 15% per yr.
b. The DCF internal rate of return over the life of the process.

4.10. *Source Reduction-10* (source reduction, pollution prevention, domestic applications, diapers). [ap] Consider two types of diapers: commercially laundered cloth diapers and disposable, superabsorent diapers, which are known to need changing less often than cloth diapers.

a. Test your intuition on the life cycle environmental impacts of the two types of diapers with respect to the following criteria:
   - net energy requirements
   - atmospheric emissions
   - industrial and postconsumer solid waste
   - wastewater
   - water volume requirements

b. Study the report, "Energy and Environmental Profile Analysis of Children's Disposable and Cloth Diapers," by Franklin Associates, Ltd., prepared for the American Paper Institute's Diaper Manufacturers' Group, Prairie View, Kansas, 1990. Check your answers in Part a.

4.11. *Source Reduction-11* (source reduction, basic principles, mass balances). [jt] A steel machining line uses waste hydraulic oil as a lubricant for operations such as punching, drilling, and welding. The waste oil is removed from the machined parts in a vapor degreaser. The solvent, 1,1,1-trichloroethane (TCA), is manifested as hazardous waste. You are asked to estimate emissions from the degreasing process. Data available include the annual TCA input, oil removed from the parts, and spent TCA/oil solution composition, as shown in the figure below: (jt)

4.11. *Source Reduction-11 (continued)*

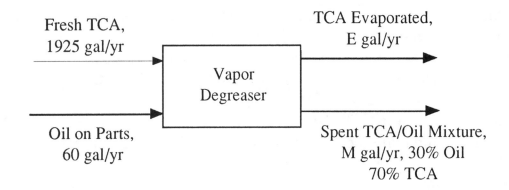

a. Calculate the annual TCA evaporative losses (gal/yr).
b. One process modification can reduce evaporative losses by 50%. Calculate the money saved on TCA purchase costs by implementing this modification.
c. Solvent recovery can be used to recycle TCA from the spent TCA/oil mixture. Calculate the annual savings in TCA purchase cost if 80% of the spent TCA is recovered.

Note that in Part c, the oil could also be recovered and re-used as a lubricant. Of course, capital and other operating costs must be considered when evaluating process modifications.

Data: ρ_{TCA} = 134 g/cm^3, TCA cost $0.45/lb

4.12. *Source Reduction-12* (source reduction, recycling, energy recovery, municipal solid waste). [wrl] Municipal solid waste (MSW) recycling is often described as the ultimate solution to the waste crisis in the U.S. It is touted to have many benefits, including volume reduction and energy and materials recovery. In practice, it is difficult to simultaneously optimize all of these potential benefits. In the Recycling Problem 5.8, the data provided in Table 1 were used to determine the optimum recycling rates for volume reduction and minimum tipping fees. Given the data in the Tables 1 and 2 below, rank the priority of recycling of the waste components to maximize each of the following goals:

1 - Volume reduction
2 - Weight reduction
3 - Energy recovery and volume reduction (no incineration allowed)
4 - Energy recovery and volume reduction (incineration allowed)

4.12. *Source Reduction-12 (continued)*

5 - Energy recovery and weight reduction (no incineration allowed)
6 - Energy recovery and weight reduction (incineration allowed)
7 - Total Value of recovered materials

Table 1. Municipal solid waste volume and weight composition for specific waste components.

Component	Volume %	Weight %
Paper/Paperboard	34.1	34.2
Glass	2.0	7.1
Aluminum	2.3	1.1
Recyclable Steel	9.8	7.0
Plastics	19.9	9.2
Yard Waste[1]	10.3	19.8
Other[2]	21.6	21.6

1 - Not recyclable but compostable; 2 - Not recyclable.

Table 2. Municipal solid waste energy and market value data for specific waste components.

Component	Energy Savings from Recycling[1] Btu/lb	Energy Savings from Recycling[1] Btu/ft³	Market Value $/lb	Market Value $/ft³
Paper	3,008	83,555	-0.0025	-0.0694
Glass	2,578	57,289	0.0075	0.1667
Aluminum	95,387	171,643	0.26	0.4678
Steel	5,450	30,277	0.01	0.0555
Mixed Plastics[2]	14,000	19,704	nil	nil
Mixed Plastics[3]	nil	nil	nil	nil
Yard Waste[4]	nil	nil	nil	nil
Other[5]	nil	nil	nil	nil

1 - Does not include post-consumer sorting or transportation costs; 2 - Incineration; 3 - Non-incineration; 4 - Not recyclable but compostable; 5 - Not recyclable.

4.13. *Source Reduction-13* (source reduction, mass balance, carbon dioxide, recovery). [nb] Recently attention has been focussed on CO_2 emissions because it is believed (theoretically) to contribute to more than half of the global warming problem. The sources of CO_2 emissions in the United States are summarized in the pie chart below.

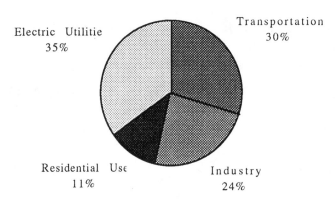

Source of U. S. Carbon Dioxide Emissions

Electric utilities produce a third of all U. S. CO_2 emissions, and are ideally situated to improve conversion efficiency as well as to promote efficiency of energy use by the public. In a thermal power plant, heat energy generated by the combustion of fossil fuels is converted to electricity. Combustion of fossil fuels (coal, natural gas and fuel oils) may be represented by the following equation:

$$CH_x + (1 + 0.25_x)\, O_2 \rightarrow CO_2 + 0.5_x\, H_2O \qquad (1)$$

where CH_x and O_2 represent the fossil fuel and stoichiometric oxygen, respectively. For coal, x varies depending on its hydrogen content; for natural gas, the value of x is approximately 4. For simplification, coal and natural gas can be considered to be carbon (x = 0) and methane (x = 4), respectively. Thus, combustion of coal and natural gas can be represented by:

$$C + O_2 \rightarrow CO_2 \qquad (2)$$

$$CH_4 + 2\, O_2 \rightarrow CO_2 + 2\, H_2O \qquad (3)$$

a. Oxygen consumed in the combustion of coal and natural gas is provided solely by air. Determine the number of gmol of N_2 generated in the flue gas (all combustion gases) per gmol of O_2 burned. Assume the composition of the air to be 79 mol% N_2 and 21 mol% O_2.

4.13. *Source Reduction-13 (continued)*

b. What is the percentage of CO_2 in the flue gas if 20% excess air is used in the combustion of (i) C and (ii) CH_4?

c. The efficiency of thermal energy conversion is indicated by the heat rate, which is 14,000 BTU/kW–h of energy generated. The amount of heat released by combustion of coal (assumed to be C) is represented by the lower heating value (LHV) and is 14,000 BTU/lb C. Calculate the moles of flue gas generated by a 200 MW electric power plant operating with 20% excess air.

d. Recovery of CO_2 is commonly accomplished using a monoethanol-amine (MEA) absorber shown schematically in the figure below. The chemistry behind the absorption and stripping processes may be represented by the following equation:

$$2 \ (HOC_2H_4)NH_2 + CO_2 + H_2O \ \rightarrow \ (HOC_2H_4NH_3)_2CO_3 \qquad (4)$$

At high temperatures (>200°F), the reaction reverses to the left. If the capital and operating cost (over a 10-year lifetime) is estimated at $40,000,000, what is the estimated cost of recovering one gmol of CO_2 gas generated? Express cost in $/gmol of CO_2 and in $/KW-h generated, and assume the plant operates 365 d/yr. Ignore all interest and inflation costs.

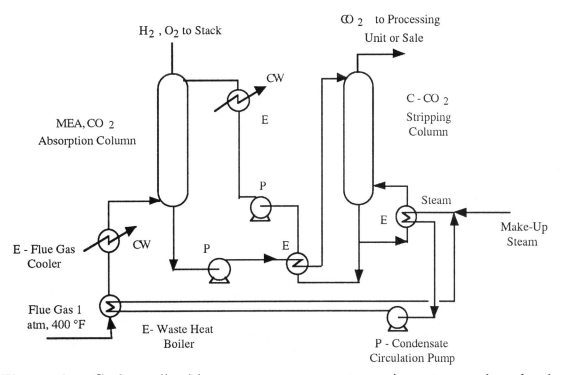

Figure 1. Carbon dioxide gas recovery system via a monoethanolamine (MEA) absorber.

5. Recycling

5.1. *Recycling-1* (pyrolysis, ethane, recycling). [lar] Ethane is converted to ethylene, a valuable monomer, via pyrolysis. The one-pass conversion efficiency of ethane ranges from 40 to 70% depending on operating conditions of a particular facility, but can be easily set to a desired value by controlling the temperature of the reaction. If ethane is not recycled within the plant it must be sent to a flare and be wasted. The flow schematic for the process is shown below.

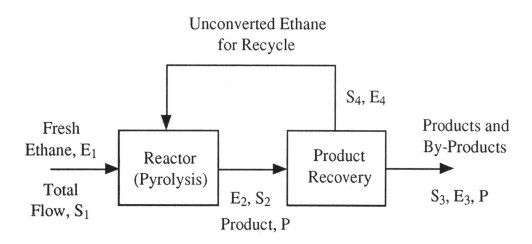

a. For a fresh ethane feed rate of 100,000 kg/h, evaluate the impact of one-pass ethane conversion efficiency on the plant recycle rate and hydraulic loading through carrying out a material balance on the process and tabulating recycling flows for 40, 50, 60 and 70% conversion efficiency. For your calculations assume that no ethane leaves the Product Recovery unit, i.e., $E_3 = 0$.

b. Given the following additional data, what one-pass conversion efficiency would you recommend, and why? Ethylene Yield (Y mass ethylene/mass ethane converted) is related to one-pass ethane conversion (C, %) by the following expression:

$$Y = 0.85 + (40 - C)/300$$

5.2. *Recycling-2* (economics, life-cycle costs, waste disposal, recycling). [lar] As Director of Environmental, Health and Safety for Biguns Chemical Company, you are evaluating the life-cycle costs of disposing of a waste material. Your process engineering department has estimated that the initial capital investment of pollution prevention equipment required to

5.2 *Recycling-2 (continued)*.

recycle this waste material is $125,000.   Your own staff tells you that the current cost of waste disposal, including future liabilities is $60,000/yr. This disposal/future liability costs would be reduced to $25,000/yr if the waste material is recycled.   Biguns' internal rate of return for all capital investments is set at 15%.

   a. Assuming that you expect the product to be made for only five more years in its present form, do you recommend the recycling process modification?
   b. Would your recommendation be any different if Biguns expected to manufacture the product for 10 more years?
   c. What is the rate of return for the capital investment if the product is to be manufactured for 10 more years?
   d. What would your recommendation be if the annual present worth of expenses for the two options were as shown below:

Year	Costs, $ No Recycling	Costs, $ With Recycling
1	64,321	22,492 + 125,000
2	65,356	23,451
3	68,879	25,012
4	71,034	27,914
5	72,308	29,037

NOTE: The classic formula that represents the time value of money is:

$$F = P (1 + i)^n$$

where P = present worth or costs, $; F = future worth or costs, $; n = number of discount periods, usually in years; and i = interest rate per discount period, %.

The total cost for any project can be determined from the sum of the initial investment costs, C_o, plus the present value of all of the future costs, F_n, for each period, n, using the following expression:

$$P = C_o + \Sigma \left[\frac{F_n}{(1 + i)^n} \right]$$

5.3. *Recycling-3* (potassium nitrate, evaporator, crystallization). [cr] The manufacture of potassium nitrate involves its recovery from an aqueous solution of 20% KNO_3. During the process, the water is evaporated, leaving an outlet stream with a concentration of 50% KNO_3. This stream is fed into a crystallization unit which generates an outlet product of 96% KNO_3 (anhydrous crystals) and 4% water. A residual aqueous solution, which contains 0.55 g of KNO_3 per g of water, also leaves the crystallizer and is mixed with the fresh solution of KNO_3 entering the evaporator. Such recycling minimizes raw materials losses and associated pollution.

a. Draw a schematic to represent the process as described above.
b. What is the mass flowrate (kg/h) of the recycled material (R) when the feedstock flowrate (F) is 5000 kg/h?
c. What is the effect of the concentration of KNO_3 in the residual water to the crystallizer on the recycle flowrate? Show this effect by plotting recycle flow (R) versus crystallizer feedstock concentration (M).

5.4. *Recycling-4* (degreasing, metal cleaning, TCA, CFCs, vapor recovery). [dks] Metal surfaces are often cleaned using organic solvents in an open top degreasing tank. One of the widely used solvents for such operations is 1,1,1–trichloroethane (TCA). TCA belongs to a group of highly stable chemicals known as ozone depleters. A typical degreasing operation is depicted in Figure 1. The emission factor for the process shown is estimated at 0.6 lb/lb of TCA entering the degreaser. The solvent from the degreaser is sent to a solvent recovery unit where 80% of the solvent is recovered and 20% of the solvent is disposed of with the sludge.

a. In order to assess the desirability of installing of a vapor recovery system, determine the mass of TCA vented to the atmosphere/lb of fresh TCA used.
b. By varying the emission factor from 0.3 to 0.9, examine its impact on the feed rate to the solvent recovery unit.

5.4 Recycling-4 (continued).

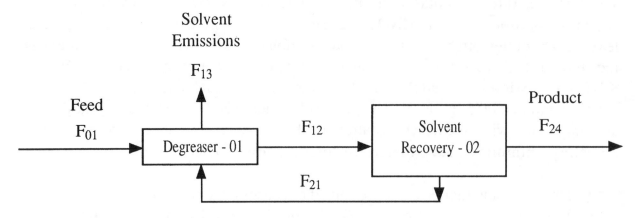

Figure 1.   Schematic of a typical degreasing operation.

5.5. Recycling-5 (degreasing, metal cleaning, TCA, CFCs, vapor recovery).   [dks]   In Problem 5.4, a typical degreasing operation was described.   In this problem, an evaluation is made of a degreasing operation that utilizes a vapor recovery system on the degreasing unit (Figure 1).   The emission factor for the process shown is estimated at 0.6 lb/lb of TCA entering the degreaser.   The solvent from the degreaser is sent to a solvent recovery unit where 80% of the solvent is recovered and 20% of the solvent is disposed of with the sludge.   The vapor recovery unit is 90% efficient.

a. Determine the mass of TCA vented to the atmosphere/lb of fresh TCA used.
b. By varying the emission factor from 0.3 to 0.9, evaluate its impact on the feed rate to the solvent recovery unit.
c. Compare these results to those obtained in Problem 5.4 above.

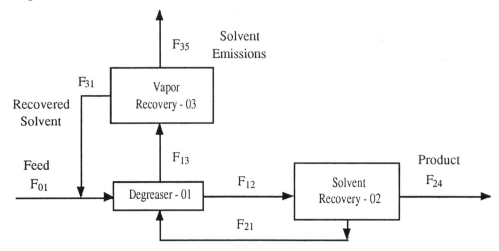

Figure 2.   Schematic of degreasing operation with vapor recovery unit.

5.6. *Recycling-6* (conservation of mass; material balances). [rrd] Perchloroethylene (PCE) is utilized in a degreasing operation and is lost from the process via evaporation from the degreasing tank. This degreasing process has an emission factor (estimated emission rate/unit measure of production) of 0.78 lb PCE released/lb PCE entering the degreasing operation. The PCE entering the degreaser is made up of recycled PCE from a solvent recovery operation plus fresh PCE make-up. The solvent recovery system is 75% efficient, with the 25% reject going off-site for disposal (Adopted from EPA 560/4-88-002, December 1987, "Estimating Releases and Waste Treatment Efficiencies for the Toxic Chemical Release Inventory Form," Office of Pesticides and Toxic Substances, Washington, D.C.).

The process diagram for the degreasing/solvent recovery system is shown below:

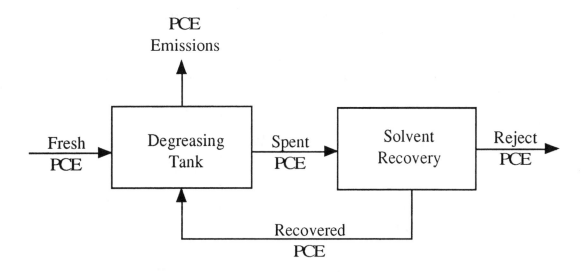

a. Develop a mass balance around the degreaser.
b. Develop a mass balance around the solvent recovery system.
c. Develop a mass balance around the entire system.
d. Determine the mass of PCE emitted per pound of fresh PCE utilized.
e. What is the impact of the emission factor in the degreasing operation on the flow rates within the solvent recovery unit? Please quantify.

5.7. *Recycling-7.* (Life Cycle Analysis, energy conservation, atmospheric emissions, product manufacturing). [rl] Life Cycle Analysis (LCA) is a technique often used to determine the impact of recycling on a full range of operations and processes. LCA of alternative products can be

5.7 *Recycling-7 (continued).*

summarized with parameters representing energy requirements of and atmospheric emissions from production of each item.

Two alternative products are being considered for manufacture by a company.   The following four parameters are being utilized to describe energy requirements and air emission rates for each product as they begin their LCA process:

E1 = energy requirements for raw material acquisition and product
        disposal,
E2 = energy requirements for material processing, product manufacture
        and product use,
G1 = atmospheric emissions from raw material acquisition and product
        disposal,
G2 = atmospheric emissions from material processing, product
        manufacture and product use.

The two products have the following parameter values for use in the LCA:

	Product 1	Product 2
E1	185 J/bag	724 J/bag
E2	464 J/bag	905 J/bag
G1	0.414 g/bag	1.463 g/bag
G2	1.000 g/bag	1.478 g/bag

a. What are the energy requirements and total emissions for each
   product?
b. Suppose that 25% of each product is recycled. Find the total energy
   requirement and total emissions per unit of each product.
c. Repeat Part b for a 50% recycle rate.
d. If the products may be substituted at the rate of one unit of Product
   2 = 2 units of Product 1, find the recycle rate at which energy
   requirements for the two products are identical.

5.8. *Recycling-8.*   (municipal solid waste, resource recovery, economics).   [wrl]   You are the engineer responsible for your municipality's solid waste disposal.   The legislature has passed a law mandating that all municipalities have a municipal solid waste (MSW)

5.8 Recycling-8 (continued).

recycling plan that will reduce the solid waste disposal rate by 25%. Your city's landfill is being operated under contract with a disposal company which charges a tipping (disposal) fee of $65/ton.

Landfill tipping fees are almost always weight-based charges (as is done above), however, landfill space is filled by volume, not weight. If the city owns the landfill, it will be concerned about how quickly the landfill space is being used up. On the other hand, if the city landfill is owned by others, the city will be more concerned with the direct cost of disposal. Some statewide environmental regulatory agencies have adopted rules to give municipalities an incentive to conserve landfill space.

a. Determine the percentage of MSW component (or components) that must be recycled to achieve a 25% volume reduction and the minimum tipping fee after recycling. Assume no recycling at the present time, and use the data for the composition of the waste stream given below:

Component	Volume %	Weight %
Paper/Paperboard	34.1	34.2
Glass	2.0	7.1
Aluminum	2.3	1.1
Recyclable Steel	9.8	7.0
Plastics	19.9	9.2
Yard Wastes[1]	10.3	19.8
All Other[2]	21.6	21.6

b. Assuming the following values for recycled materials, would your answer to Part a be any different? Please comment.

Component	Value, $/ton
Paper/Paperboard	25
Separated Glass	100
Mixed Glass	40
Aluminum	30
Recyclable Steel	100
Separated Plastics	200
Mixed Plastics	50
Yard Wastes[1]	0
All Other[2]	0

1.   Compostable.
2.   Non-recyclable.

5.9. *Recycling-9*. (basic concepts, glass recycle). [ihf] In the EPA hierarchy of pollution prevention, source reduction and recycling are the most viable techniques. Glass food and beverage containers are excellent candidates for recycling as these can be used many times. However, there are some glass items that should not be recycled.

a. Suggest a few containers which should not be recycled, and why they should not be recycled.
b. Does a similar problem exist for plastics, limiting their ability and desireability to be recycled?

5.10. *Recycling-10*. (basic concepts, emission rates, ideal gas law). [ihf] The first step in pollution prevention is to estimate the concentration of the pollutant that is being controlled. A large laboratory with a volume of 1100 m^3, at 22°C and 1 atm, contains a reactor which may emit as much as 1.50 gmol of a hydrocarbon (HC) into the room if a seal ruptures. If the hydrocarbon mole fraction in the room air becomes greater than 850 ppb, it constitutes a health hazard.

a. Suppose the reactor seal ruptures and the maximum amount of HC is emitted almost instantaneously. Assume that the air flow in the room is sufficient to make the room behave like a continuous stirred tank reactor (CSTR), i.e., the air composition is spatially uniform. Calculate the ppb of hydrocarbon in the room. Is there a health risk?
b. What might be done to decrease the environmentally hazardous nature of the reactor?
c. What might be done to improve pollution prevention?

6. Treatment/Disposal

6.1. *Treatment/Disposal-1* (disposal, hazardous waste, landfill, design considerations). [ap] Describe technical issues that must be considered in the design of a hazardous waste landfill.

6.2 *Treatment/Disposal-2* (treatment, risk analysis, transportation, liability, regulation). [wrl] In evaluating the economic impacts of pollution prevention strategies, costs of alternative treatment options must be considered. In some cases, the reduction of potential liability alone may be adequate to justify capital expenditures for pollution prevention. The estimation of the clean-up cost of a spill or release is also an important factor. The following problem demonstrates a practical method of estimating transportation liability.

Biguns Chemical Co. transports solid hazardous waste to a disposal site. On average, Biguns' hauling trucks carry 4 tons of waste/trip for a total of 32,000 tons/yr. In the event of a truck overturn, it can be assumed that 2 tons of the waste is spilled. DOT statistics indicate that 1 out of 4,000 waste hauling trucks overturns during an average trip. Industry studies and Chevron Corp.'s SMART (Save Money And Reduce Toxics) data indicate that cleanups resulting from transportation spills cost as much as $10,000 per ton. Calculate the total annual potential liability of producing and "disposing" of the waste in this matter. Express the answer in $/yr and $/ton.

6.3. *Treatment/Disposal-3* (combustion, hydrocarbons, stoichiometry). [cr] The engineering division of a company has made a decision that it would be worthwhile to combust a pure, dilute hydrocarbon gas stream and recover the heat generated from this combustion reaction as a make-up heat source for a separation unit elsewhere in their plant.

Determine the chemical formula for this compound if the flue gas composition on a bone dry basis (by gmol or volume) is:

CO_2: 7.5%, CO: 1.3%, O_2: 8.1%, N_2: 83.1%

It is known that this hydrocarbon is not oxygenated.

6.4. *Treatment/Disposal-4* (tannery waste, filter press, sludge dewatering, chromium sludge). [kg] A tannery uses a filter press to dewater its raw sludge. The dewatered sludge contains 120 mg of total chromium/kg of sludge. The filter press has a surface area of 150 ft^2 with a filtrate rate of 10 gal/h-ft^2. Assume that it is 100% efficient in separating solids. Estimate the amount of total chromium disposed/yr from this plant if the plant operates 16 h/d, and 7 d/wk, 50 wk/yr. The following information is available to you.

Before dewatering:
Water content of sludge = 95%; Solid content of sludge = 5 wt%

After Dewatering:
Water content = 52%; Solid content = 48 wt%

6.5. *Treatment/Disposal-5* (treatment, laboratory chemicals, economics, sulfides, stoichiometry, conversions). [glh] If source reduction and recycling are not feasible options for the prevention of pollution, the treatment of this waste should be considered. In many cases the treated chemicals are rendered innocuous and can be disposed of through a biological or physical/chemical waste treatment plant. On the other hand, if the substance cannot be rendered harmless, then such treatment is usually followed by the fourth (and least desirable) option, landfill disposal.

Of the many treatment methods, oxidation is frequently used when inexpensive oxidants are available. Some of these include air, hydrogen peroxide and common bleach, sodium hypochlorite. Ozone, though expensive, is occasionally used in wastewater treatment. In the disposal of any chemical in any amount, local laws and regulations relating to such disposal must be complied with.

Ammonium sulfide is a noxious substance which can be used for the generation of sulfide ion in inorganic qualitative analysis. The substance is also used in the photographic and textile industries. Its aqueous solution has a high pH ($\approx$ 11 to 12), has a foul odor, and reacts with acids to generate hydrogen sulfide, a toxic and foul-smelling gas. It cannot be disposed of directly but must be treated.

6.5. *Treatment/Disposal-5* *(continued).*

Fortunately, ammonium sulfide can be oxidized quite easily with a hypochlorite solution (Lunn and Sansone, 1990), such as household bleach, according to the following equation:

$$(NH_4)_2S + NaOCl \rightarrow NH_4Cl + Na_2SO_4$$

a. Balance the above equation.
b. Calculate the cost of disposal of one gallon of ammonium sulfate solution using household bleach.
c. Compare the economics of this disposal by oxidation with the cost of simply calling in a waste disposal company to place it in a "lab pack" and take it away.

Data:

One container of aqueous $(NH_4)_2S$ weighs 2.3 kg. The solution is 20.1% $(NH_4)_2S$.

Household bleach, NaOCl, contains 5.25% active ingredient, weighs about 8.7 lb/gal and costs $0.89 (retail).

One "lab pack" (55 gal drum) will hold about 12 containers of aqueous ammonium sulfide and costs $350 to dispose of through a qualified disposal company.

Lunn, G. and E. B. Sansone. 1990. *Destruction of Hazardous Chemicals in the Laboratory.* John Wiley and Sons, Inc., New York.

6.6. *Treatment/Disposal-6* (hydrocarbons, biodegradation, ground water, hydrogen peroxide). [dem] In-situ biodegradation is a feasible remediation technology for hydrocarbons trapped in the soil. Microorganisms can effectively use the hydrocarbons as substrate (food) and degrade them to carbon dioxide and water under aerobic conditions, provided that there are micronutrients (such as nitrogen and phosphorous) and sufficient oxygen available to them. One method of introducing oxygen to the system is by injecting hydrogen peroxide into the soil.

a. Estimate a range of the oxygen/hydrocarbon mass ratio required to biodegrade straight-chain aliphatic hydrocarbons.

6.6. *Treatment/Disposal-6   (continued).*

b. Assuming that 3.5 kg of oxygen are needed to degrade 1 kg of hydrocarbon, estimate the cost of providing oxygen to degrade 5,000 gallons of gasoline (with a density of 700 kg/m^3) if hydrogen peroxide costs \$2.20/kg. Neglect volatilization, and assume 100% utilization of the applied hydrogen peroxide.

6.7. *Treatment/Disposal-6* (ammonia, absorption, packed bed tower). [cr] An absorption tower is used to remove the ammonia contained in a gaseous mixture of ammonia in nitrogen. The volume (gmol) fraction of ammonia is 12%. Water is fed from the head of the column toward a packed bed where the upward flow of the gaseous mixture is mixed with the water stream. Under steady-state conditions, 90% of the ammonia in the gas is recovered in the aqueous phase. The mass fraction of ammonia in the liquid stream leaving the absorber is 0.04. Nitrogen is virtually water insoluble.

Calculate the number of liters of the aqueous phase produced after treating 100 gmol of the gaseous mixture. The specific gravity of the solution is 0.98.

6.8. *Treatment/Disposal-8* (air stripper, TCE contaminated ground water). [dem] A packed-tower air stripper is used to treat 30 ppm of TCE at 20°C from contaminated ground water. A blower is used to pass the air into the packed tower. The stripper was designed such that the height of the packing is 4 m, the diameter is 1.5 m, and the height of a transfer unit is 1 m.

a. If the stripping factor used in design is R=5, what is the removal efficiency?
b. If the blower produces a maximum air flow of 3 standard m^3/min, what is the maximum water flow that can be treated by the stripper?

Assume a Henry's constant of 324 atm at 20°C and employ the following equations:

6.8. *Treatment/Disposal-8   (continued).*

$$Z = (NTU)(HTU)$$

$$NTU = \frac{R}{R-1} \ln\left[\frac{(C_{in}/C_{out})(R-1)+1}{R}\right]$$

$$R = \frac{H}{P_t}\frac{G}{L}$$

$$HTU = \frac{L}{(K_L a)(C_0)}$$

where Z = height of the packing, m; NTU = Number of transfer units; HTU = Height of a transfer unit, m; R = stripping factor; C_{in} = Inflow liquid concentration, ppm; C_{out} = Outflow liquid concentration, ppm; H = Henry's constant, atm; P_t = atmospheric pressure, 1 atm; G = air loading rate, kgmol/s-m^2; L = water loading rate, kgmol/s-m^2; $K_L a$ = packing coefficient, 1/s; and, C_0 = molar density of water, 55.6 kmol/m^3 at 20°C.

At standard conditions, the molar air-water ratio, G/L, is related to the volume air-water ratio through $G''/L'' = 1325\ G/L$, where the units of G'' and L'' are m^3/s-m^2.

Reference: Noonan, D., and J. Curtis (1990), *Ground Water Remediation and Petroleum: A Guide for Underground Storage Tanks*, Lewis Publishers, Boca Raton, FL.

6.9. *Treatment/Disposal-9* (air stripper, BTEX contaminated ground water, iron removal). [dem] An air stripper is used to treat 30 gpm of ground water contaminated with 25 ppm of BTEX (benzene, toluene, ethylbenzene, and xylenes) to a final concentration of 100 ppb. The tower is 4 m high and 0.6 m in diameter, while the porosity of the packing is 0.56.

The project is located in the Northeastern United States where ground water naturally contains 35 ppm of iron as Fe^{2+}. The iron precipitates as Fe_2O_3, filling the pores in the packed tower, which results in clogging and/or fouling. Approximately 20% of the sludge can be flushed out of the stripper during normal operations. The Fe_2O_3 sludge density is 3,200 kg/m^3. Experience indicates that a 10% pore volume reduction in the stripper decreases its efficiency by 5%.

6.9. *Treatment/Disposal-9* (continued).

How often should the stripper be taken off-line for cleaning in order to realize no more than a 5% reduction in efficiency on an on-going basis?

6.10. *Treatment/Disposal-10* (brewery waste, neutralization, equilibrium constants). [jt] Brewery vessels are cleaned using concentrated sodium hydroxide or sodium silicate solutions to dissolve organic matter. This is followed by a water rinse, yielding a dilute alkaline solution. As this is a periodic step, the brewery wishes to neutralize the waste rinsewater to avoid pH upsets at the Publicly Owned Treatment Works (POTW). One approach is to absorb gaseous CO_2 vented during fermentation; this not only neutralizes the wastewater but also reduces emissions of CO_2, a greenhouse gas.

   a. Calculate the mass of CO_2 required to neutralize the rinsewater from the production of one bbl (barrel, U.S. liquid) of beer. Assume:
- initial rinsewater pH = 12.0; solution is NaOH
- rinsewater pH after CO_2 absorption = 7.0
- (bbl rinsewater)/(bbl beer produced) = 5
  (Note: Neutralization is only to be carried out for the rinse water; the total amount of water sent to the POTW is significantly greater)
- T = 25°C
- CO_2 absorption efficiency = 50%, were "absorption efficiency" is defined as the percent of the CO_2 absorbed by the rinsewater.

     Remember to account for carbonate equilibria, where:

$$k_{A1} = \frac{[H^+][HCO_3^-]}{[H_2CO_3]} = 4.30 \times 10^{-7} \, M @ 25°C$$

$$k_{A2} = \frac{[H^+][CO_3^{-2}]}{[HCO_3^-]} = 5.61 \times 10^{-11} \, M @ 25°C$$

   b. CO_2 is vented from the fermentor at a rate of 9 lb CO_2/bbl beer produced. What fraction must be collected to neutralize the rinsewater?

   c. Fermentation and cleaning are batch operations with very different time scales and scheduling. How will this impact process design?

6.11. *Treatment/Disposal-11* (basic principles, soil remediation, DNAPL, solubility, dissolution). [dem] A Dense Non-Aqueous Phase Liquid (DNAPL) is a low-solubility compound with density greater than water; an example is trichloroethylene (TCE). DNAPLs are also called "sinkers" since they drop to the bottom of an aquifer upon contact with the water table. Once in the soil or the ground water, DNAPLs are extremely difficult to remove.

Assume that a TCE spill has contaminated an aquifer where the specific discharge (ground water flow per unit area) is 0.02 m³/m²-d. Assume a cube of soil, 1 m on each side, where the ground water flows normal to one of the faces (see the figure below). The porosity of the soil is 0.35 and the residual saturation (the fraction of the void space occupied by the fluid) initially is 20% for TCE and 80% for water.

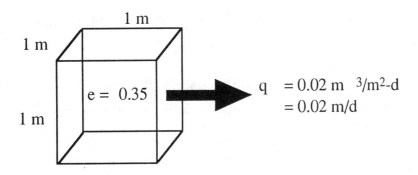

If the proposed removal mechanism is solely by dissolution, find:

a. The number of pore volumes required to remove all the TCE. Assume that all water passing through the cube is available for dissolution of the TCE.
b. The time to remove all of the TCE.

The density of TCE is 1,470 kg/m³ and its solubility in water is 1.1 g/L.

6.12. *Treatment/Disposal-12* (activated carbon adsorption, water treatment, Freundlich isotherm, treatment costs). [dem] Granular activated carbon (GAC) adsorption has been selected as the technology to clean 10 gpm of water contaminated with 5 ppm of benzene. A laboratory test yields the following sorption isotherm results:

6.12. *Treatment/Disposal-12* *(continued).*

Solution Concentration (ppm)	Adsorbed weight/Carbon weight (mg/g)
4	8.5
5.7	11.5
6.1	22
10	50
12	57

a. Determine the Freundlich sorption isotherm for benzene if the Freundlich isotherm is generated from a logarithmic linear regression of sorption data according to the following expression:

$$X/M = a\ C^b$$

where X = mass of adsorbate (benzene), mg; M = mass of adsorbent (GAC), g; C = Concentration of benzene in water, ppm; and, a and b are regression constants.

b. If the cost of carbon is $7/lb, what is the annual treatment cost if the adsorber is operated to its equilibrium capacity, not to breakthrough?

6.13. *Treatment/Disposal-9* (biofilters, gas treatment, kinetics, ethyl acetate). [dks] Organic compounds may be effectively removed from gas streams using biofilters. In a pilot plant study, five packed columns, consisting of peat moss as packing material, were placed in series. The following experimental data are available for the removal of ethyl acetate from a gaseous air/ethyl acetate mixture:

Column #	Exit gas concentration (g/m³)
1	0.30
2	0.25
3	0.15
4	0.10
5	0.01

The following additional data are available:
Inlet gas concentration = 0.4 g/m³, Gas velocity = 3 cm/s; Column Height = 60 cm

6.13. *Treatment/Disposal-13* *(continued)*.

Assuming zero order reaction kinetics, determine the required surface area for 90% removal of ethyl acetate in a commercial unit with the following operating conditions:

Gas flow rate = 15,000 m^3/h; Packing height = 1 m; Inlet concentration: 0.5 g/m^3

6.14. *Treatment/Disposal-14* (aerosol emission, flue gas treatment, scrubbers, sewage sludge incineration). [nb] Aerosols can be defined as solid and liquid particles suspended in a gas. They are produced in many industrial and natural processes, and can be a very visible source of pollution (e.g., the Los Angeles smog). Flue gases are released into the atmosphere from high temperature processes, such as steel refining, industrial and domestic waste incineration, petrochemical operations, and coal–fired power plants. This problem examines the reduction of aerosol emissions in flue gases, using data obtained from municipal waste incineration. The principles are also applicable to other aerosol– producing processes.

Municipal sewage sludges often consist of sewage mixed with industrial and domestic chemicals. They can contain a wide variety of hazardous and toxic substances. Incineration is often a preferred method for sludge treatment because the volume of waste is greatly reduced and hazardous organic compounds are destroyed. In some situations, energy can also be recovered. Mechanisms for particle formation in coal combustion are shown in Figure 1, and similar mechanisms operate in waste incineration. Gas cleaning equipment, such as scrubbers, electrostatic precipitators, and baghouses, remove the bulk of these particles before the flue gas is released into the atmosphere.

The size distribution of aerosols is important because inhaled particles can be hazardous depending on particle size and composition. Particles smaller than about 10 μm diameter (1 μm = 10^{-6} m) can be retained in the lungs. Particles between 2 and 10 μm diameter can penetrate deep into the alveolar region, where gas exchange occurs (Hinds, 1982, Chapter 11). Soluble particles can be absorbed through the lungs into the bloodstream, and some particles (e.g., some forms of asbestos) can cause lung disease.

6.14. *Treatment/Disposal-14 (continued).*

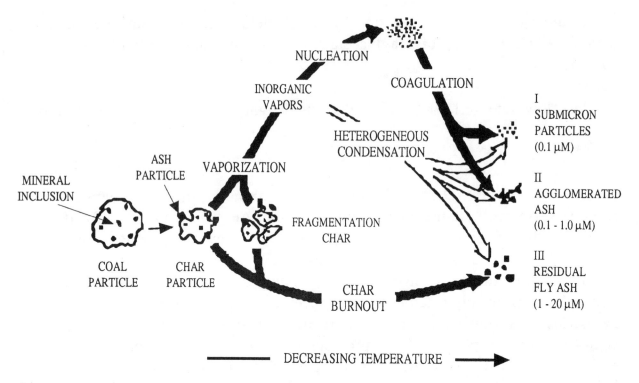

Figure 1.   Schematic of ash formation processes during coal combustion.

Measurements were made of particles emitted from a municipal sludge incinerator.   Samples were taken from the flue gas entering and leaving the scrubber.   The data are summarized in Table 1.   The mass of particles in each size interval was measured on a basis of 1 m^3 of flue gas at 298 K and 1 atm.

Table 1: Size distribution data (adapted from Bennett and Knapp, 1982)

Size Interval (Δd_p) μm	Entering Scrubber Particle mass (Δm) mg/m^3	Leaving Scrubber Particle mass (Δm) mg/m^3
0.15 – 0.34	554	9.5
0.3 - 0.5	18.4	10.1
0.5 - 1.0	25.1	7.2
1.0 - 1.8	30.1	4.5
1.8 - 3.8	155	4.1
3.8 - 9.6	500	4.4
9.6 - 22.0	908	4.1
22.0 - 50.0	1212	2.5

Note: Discharge gas flow rate has been adjusted to 298 K and 1 atm.

6.14. *Treatment/Disposal-14   (continued).*

a. Particle size distribution is frequently reported as a mass distribution curve (Hinds, 1982, Chapter 4). It is often plotted as a histogram, where the width of each rectangle represents the size (diameter) interval, and the height is the mass or mass fraction of particles within that interval. Since the mass of material in an interval is dependent on the width of the size interval, the mass in each interval is normalized by the interval width.

Plot the mass fraction distributions for aerosols entering and leaving the scrubber. Use a logarithmic scale on the abscissa and use dimensions of μm for diameter. Normalize the mass fraction of each interval using the width of the logarithm of particle size:

$$\Delta \log_{10} d_p = \log_{10} d_{upper} - \log_{10} d_{lower} \qquad (1)$$

where d_{upper} is the particle diameter at the upper limit of the interval, and d_{lower} is the diameter at the lower limit.

b. Explain the shape of the size distribution of the aerosol entering the scrubber. (Refer to Figure 1 to relate particle sizes to specific formation processes).

c. The mass mean diameter is the arithmetic average of the mass distribution. It is defined as:

$$d_m = \frac{\sum_i m_i d_i}{M} \qquad (2)$$

where m_i is the mass of particles in size range i, d_i is the geometric average diameter of size range i

$$d_i = \sqrt{d_{upper} d_{lower}}$$

and M is the total mass of all size ranges.

Calculate the mass mean diameters (d_m) of particles entering and leaving the scrubber. Compare the size distributions and mean diameters and comment on the effect of the scrubber.

6.14. *Treatment/Disposal-14   (continued).*

d. Calculate the overall efficiency ($\eta_{overall}$) of the scrubber. Total efficiency is defined as:

$$\eta_{overall} = (1 - m_{out}/m_{in})\ 100 \tag{3}$$

where m_{in} is the mass of particles entering the scrubber, and m_{out} is the mass leaving the scrubber.

e. Plot a line graph of collection efficiency (η_d) of the scrubber as a function of particle diameter (d_i).

f. Discuss the significant features of the collection efficiency graph prepared in Part e. Compare overall efficiency ($\eta_{overall}$) with the efficiency curve (η_d), and discuss the differences in the information they convey.

Question for Discussion

Apart from total emissions, what other factors should be considered when evaluating the environmental impact of particles formed by combustion and incineration?

References and Further Reading:

Bennett, R. L. and Knapp, K. T., "Characterization of Particulate Emissions from Municipal Wastewater Sludge Incinerators", *Environ. Sci. Technol.*, 16, 12, 831–836 (1982).

Brunner, C. R., *Hazardous Air Emissions from Incineration*, Chapman and Hall, New York (1985).

Helble, J. J., and Sarofim, A. F., "Factors Determining the Primary Size of Flame–Generated Inorganic Aerosols", *J. Colloid Interface Sci.*, 128, 2, 348–362 (1989).

Hinds, W. C., *Aerosol Technology*, Wiley–Interscience, New York (1982).

Liptak, B. G., *Municipal Waste Disposal in the 1990's*, Chilton, Radnor, PA (1991).

Seinfeld, J. H., *Atmospheric Chemistry and Physics of Air Pollution*, John Wiley and Sons, New York (1986).

6.15. *Treatment/Disposal-15* (aerosol emission, flue gas treatment, scrubbers, electrostatic precipitators, filters). [nb] Gas cleaning devices, such as electrostatic precipitators, filters and scrubbers, are used to remove a high proportion of aerosols formed in industrial processes. However, the particles which are removed least efficiently are also those which are potentially the most harmful to health. There are fundamental physical limitations to most conventional gas cleaning devices which result in lower collection efficiencies for particles between 0.1 and 0.6 μm diameter (Friedlander, 1977). In this problem alternative strategies for reducing and preventing aerosol emissions will be explored.

a. Particles formed by nucleation and coagulation in high temperature processes (e.g., the submicron particles in Figure 1 in the Problem 6.14) are generally agglomerates composed of many smaller primary particles (which are in the order of 1 to 10 nm diameter). Growth of these agglomerates can be described by Equation 1:

$$d_m \propto \left[T^{0.5} \, d_0^{\,3(D_f-3)/2} \, \phi t \right]^{2/(3D_f-4)} \tag{1}$$

where d_0 = the diameter of the primary particles; ϕ = particle loading, m^3 particles/m^3 of gas at 298K and 1 atm; T = temperature; and, t = residence time. D_f is a growth exponent (Matsoukas and Friedlander, 1991), which for this problem, can be considered to be a property of the incinerator. The value of D_f varies between 2 and 3.

Given a specific incinerator and assuming that D_f is fixed, assess the relative importance of each of the factors which affect agglomerate growth.

b. Discuss how you would attempt to reduce particle emissions from the waste incinerator by modifying particle size. Base your discussion on the growth equation, Equation 1.

c. Suggest other methods which may be used to modify the particle size distribution. In particular, consider ways in which kinetics of agglomeration might be enhanced.

Questions for Discussion

i. At present in the United States, combustion residues (ash) are disposed of in landfills. Suggest ways in which this material may be recycled.

6.15. *Treatment/Disposal-15 (continued)*.

ii. It is more desirable to prevent the formation of aerosols in industrial processes than to remove particles from flue gas. Suggest possible approaches which may be used in redesigning processes to prevent aerosol formation.

References and Further Reading:

Friedlander, S. K., *Smoke, Dust and Haze*, Wiley–Interscience, New York (1977).

Matsoukas, T. and Friedlander, S. K., "Dynamics of Aerosol Agglomerate Formation", *J. Colloid Interface Sci.*, 146, 2, 495–506 (1991).

Vence, T. D., "Potential of Recycling Ash from Resource Recovery Facilities in California", *Proceedings of the 1984 National Waste Processing Conference*, Orlando, Florida, ASME, 640–654 (1984).

Williams, M. M. R. and Loyalka, S. K., *Aerosol Science Theory and Practice*, Pergamon Press, Oxford (1991).

7. Chemical Plant/Domestic Applications

7.1. *Chemical Plant/Domestic Applications-1* (basic principles, equilibrium flash separator). [dlu] The equilibrium flash separator is sometimes useful as a means of separating a volatile material from a much less volatile material. It is not a very good separating device if high purities are desired. The following problem illustrates the extent of separation that can be obtained by a simple equilibrium flash separation. Ideal vapor and ideal solution behavior should be assumed.

A liquid waste stream contains 5 gmol % ethane and 95 gmol % n–hexane at 25°C. The vapor pressure of ethane at 25°C is 4150 kPa and the vapor pressure of n–hexane at 25°C is 16.1 kPa.

   a. If the pressure is such that this is a saturated liquid, what is the pressure and what will be the composition of the first vapor to form?
   b. If the pressure is reduced enough to vaporize 10 gmol% of the mixture, what fraction of ethane is removed and what is the composition of the vapor and the liquid? The temperature is held constant at 25°C.

7.2. *Chemical Plant/Domestic Applications-2* (basic principles, emission rates, ideal gas law) [ihf] The first step in pollution prevention is to estimate the concentration of the pollutant that is being generated. A large laboratory with a volume of 1100 m^3, at 22°C and 1 atm contains a reactor which may emit as much as 1.50 gmol of a hydrocarbon (HC) into the room if a seal ruptures. If the hydrocarbon mole fraction in the room air becomes greater than 850 parts per billion (ppb) it constitutes a health hazard.

   a. Suppose the reactor seal ruptures and the maximum amount of HC is emitted almost instantaneously. Assume that the air flow in the room is sufficient to make the room behave like a continuous stirred tank reactor (CSTR), i.e., the air composition is spatially uniform. Calculate the ppb of hydrocarbon in the room. Is there a health risk?
   b. What might be done to decrease the environmentally hazardous nature, or improve the safety, of the reactor?
   c. What might be done to implement pollution prevention here?

7.3. *Chemical Plant/Domestic Applications-3* (basic principles, unit conversions, gasoline). [dlu] A gasoline tank in an automobile has a 14 gal capacity. Every time the gas tank is filled, the vapor space in the tank is displaced to the environment. All forms of hydrocarbons in the atmosphere contribute to the formation of ozone and need to be controlled. This problem quantifies some of these emissions.

Assume the automobile tank vapor space, the air and the gasoline supply are all at 20°C. The vapor space is saturated with gasoline. The vapor phase mole fraction of gasoline under these conditions is about 0.4. The lost vapor has a molecular weight of about 70 g/gmol and a liquid specific gravity of 0.62. Note that gasoline is a mixture of many components so the vapor composition and molecular weight above it are a function of temperature.

  a. Calculate the amount of gasoline (as gallons of liquid) that is lost to the air during a 10 gal fill.
  b. How much is lost annually from 50 million cars filled once each week with 10 gal of gasoline?
  c. What is the value of this loss if gasoline is \$1.20/gal?

7.4. *Chemical Plant/Domestic Applications-4* (chemical reaction rates, reaction kinetics, source reduction). [jt] In some cases, it may be possible to reduce the generation of undesirable byproducts from chemical reactions by altering the operating conditions. For example, consider the reaction scheme:

$$A \xrightarrow{k_1} D$$

$$A + D \underset{k_3}{\overset{k_2}{\rightleftharpoons}} U$$

where D is the desired product and U is an undesired byproduct. Consider isothermal batch processing with no change in density. The rate equation for species A is:

$$\frac{d\,[A]}{d\,t} = -k_1\,[A] - k_2\,[A][D] + k_3\,[U]$$

where [A], [D], and [U] = concentration of A, D, and U, respectively.

7.4. *Chemical Plant/Domestic Applications-4 (continued)*.

(For additional information on chemical kinetics the reader is referred to: L. Theodore (1991), *Chemical Reactor Kinetics*, An ETS Theodore Tutorial, ETS International, Roanoke, Virginia.)

a. Develop the rate expressions for species D and U.
b. Integrate these equations for the following initial conditions (use, e.g., 4th order Runge-Kutta),

$$[A]_o = 1, [D]_o = [U]_o = 0$$

where the "o" subscript indicates initial concentrations.
Use $k_1 = k_2 = k_3 = 1$ for the rate constants. Plot $[D]/[A]_o$ and $[D]/[U]$ as a function of time for the range: $0 \leq t \leq 7$. (The ratio, $[D]/[U]$, is often referred to as the "selectivity.")
c. Study the effect of adding an inert diluent to the reactor. Using the same rate constants, integrate again and plot the results (on the same graphs produced in Part b) for:

$$[A]_o = 0.5, [D]_o = [U]_o = 0$$

$$[A]_o = 0.1, [D]_o = [U]_o = 0$$

d. For a fixed conversion of A → D, say $[D]/[A]_o = 0.85$, how does the selectivity, $([D]/[U])$, vary with $[A]_o$? Discuss these results in terms of degree of formation of undesired by-products.
e. Consider the case where a fixed production rate (mol D/yr) must be achieved. What are the trade-offs, for this reaction scheme, between reaction time, reactor volume, and selectivity? Use an 85% conversion case to demonstrate your reasoning.

7.5. *Chemical Plant/Domestic Applications-5* (open ended, material consumption, energy consumption, technology, pollution prevention, recycling). [glh] To make informed, intelligent decisions regarding pollution prevention, one must have a firm grasp of the intricate relationships between energy, technology and pollution. Pollution can often be traced back to the misuse of energy and thus one of the most potent weapons in minimizing pollution is to reduce the use of energy. Technology should be directed toward energy efficiency, i.e., toward

7.5. *Chemical Plant/Domestic Applications-5 (continued).*

getting more "bang for the Btu." The greatest strides in pollution reduction since 1970 have been achieved through energy conservation with little or no reduction in the standard of living. In fact, in most cases, large savings in cash expenditures have been achieved. It has been calculated that if the entire United States were to switch to compact fluorescent bulbs overnight, it would be transformed from an energy importing nation to an energy exporting nation.

An example calculation relates to the recycling of aluminum cans. In the United States 200 million cans are used daily but only 100 million are recycled. The remainder can be found in landfills, if not along the highway. Producing aluminum from ore is a very energy intensive process and the pollution resulting from this will be found in the carbon dioxide (greenhouse effect) generated at the power station and the mine tailings produced during the mining operations. If the wasted 100 million aluminum cans could be recycled, enough energy would be saved to power a city the size of Baltimore every day!

As a consumer society, we have been encouraged to use more material goods as a matter of course. All of these materials require energy at every step of production. Energy, regardless of its form, produces pollution in its generation and use. Below are three simple activities that we engage in everyday. Choose one of these activities and develop a "pollution tree" which traces the basic activity back to the original source of raw material, listing the type of waste/pollution at each step. You may expect to generate one main pathway and several side paths. Estimate the impact of each (large, moderate, small). Generate alternatives and or minimization techniques. Think globally!!

   a. Reading a newspaper (magazine)
   b. Washing clothes.
   c. Having a hamburger, soda and fries at the local fast food outlet.

Example: Mowing the lawn.

Mowing one lawn is a trivial activity but when the impact is multiplied by the millions of lawns in the United States, the impact may be considerable. This would be more likely to be a problem in cities such as Los Angeles and New York City which are already under a heavy pollution load.

7.5. Chemical Plant/Domestic Applications-5 (continued).

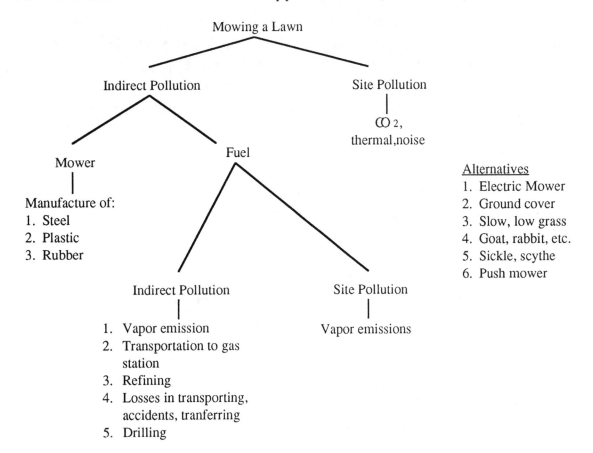

Mowing a Lawn

Indirect Pollution

Site Pollution

CO_2, thermal, noise

Mower

Fuel

Manufacture of:
1. Steel
2. Plastic
3. Rubber

Indirect Pollution
1. Vapor emission
2. Transportation to gas station
3. Refining
4. Losses in transporting, accidents, tranferring
5. Drilling

Site Pollution

Vapor emissions

Alternatives
1. Electric Mower
2. Ground cover
3. Slow, low grass
4. Goat, rabbit, etc.
5. Sickle, scythe
6. Push mower

As can be seen, each of the items under indirect pollution (manufacture, refining, etc.) will require a pollution tree of its own. Also, the use of an electric mower, of course, brings in the pollution generated at the power generating station (CO_2, other emissions) as well as the pollution caused by the fuel used at the power plant. It is up to the instructor to determine how extensive a tree is required as it quickly becomes very complex. Overall, the pollution due to lawn mowing is very small, relative to the automobile. This problem does not consider the ultimate disposal of the lawn mower nor of the clippings, if they are collected. Nor does it consider the impact of having a lawn in the first place. An extensive lawn (e.g., a golf course) which is maintained with fertilizers, herbicides and pesticides presents a very significant pollution source, especially in ground and surface water.

7.6. Chemical Plant/Domestic Applications-6 (batteries, disposal, rechargable, photovoltaics, economics). [glh] One non-point source of environmental pollution has been the introduction of heavy metals, particularly mercury and cadmium, into landfills from the casual disposal

7.6. *Chemical Plant/Domestic Applications-6 (continued).*

of household non-rechargable batteries. The major use of such batteries historically was the Type D size used in flashlights, but portable radios and tape players have made Type AA batteries much more popular. Battery manufacturers will now accept the spent batteries for recycle and/or disposal in a safe manner. Most communities have one or more collection sites - the town library, schools, etc. - where the public can turn in this hazardous waste.

A better solution to the battery problem is to switch to rechargeable batteries which can be reused over and over, lasting for years. These do require a charging device at some initial cost and a convenient 120 volt outlet for recharging. An even better system, one which is sustainable, employs the sun to do the recharging to eliminate, power company involvement altogether. Again, a special solar charger is a necessity as is, of course, a sunny day or two (Real Goods, 1992).

   a. Calculate the annual cost for the use of rechargable Nickel-Cadmium
      AA and compare it to the annual cost for conventional batteries.
      They are to be used in a portable radio which uses four such
      batteries. The rechargable batteries are recharged every month with
      a solar battery charger, which is not designed to get wet, and which
      requires two to four days to recharge the batteries depending on the
      cloud cover.
   b. What is the payback time, if there is one?

Data:

	Conventional AA	Rechargable Ni-Cad AA
Number required:	4	8
Cost (each)	$0.89	$2.75
Rotation frequency:	Monthly	Monthly
Solar recharger cost:	0	$14
Lifetime:	One Month[1]	5 yr (1,000 charges)[2]

1. Depending on intensity of use.
2. Claimed by manufacturer.

Reference:
Real Goods (1992), *Real Goods News*, Spring Issue. p.10-11. Can be
     purchased by writing to c/o Real Goods, 966 Mazzoni Street, Ukiah,
     CA 95482-3471.

7.7. *Chemical Plant/Domestic Applications-7* (open-top degreasers, RACT, regulations, VOC). [glh] Degreasing operations are a common feature of working with metal parts and tools. These usually function by dissolving grease and oil from a metallic surface, either by direct bath contact of liquid solvent or by the condensation of the volatilized vapor onto the metal parts. It is common to use a halogenated hydrocarbon for such cleaning since they are volatile, non-flammable, and are excellent solvents. However, they are toxic and if not carefully controlled, degreasing operations can be a significant source of air pollution. There are three general types of degreasers, all of which are used for either the cleaning of tools or the cleaning of metal surfaces for the next step in a production line: cold cleaners, open top degreasers, and conveyerized degreasers.

This problem focuses on open-top degreasers and the volatile organic compounds (VOCs) that may be released from them. You should review the EPA policy on Reasonable Available Control Technology (RACT) in its application to degreasing operations. Figure 1 illustrates a typical open top degreaser.

    a. Refer to Figure 1 and identify the five numbered components.
    b. In the operation of the open top degreaser, there are at least five potential vapor emission points which must be controlled. List four of them.
    c. Multiple Choice:

       1. An EPA-recommended exemption to the Control Technology Guideline (CTG) regulations is:
         a. all open-top vapor degreasers in urban nonattainment areas should be exempt from installing carbon adsorbers.
         b. open-top vapor degreasers with an open area of less than 1.0 m^2 should be exempt from installing refrigerated chillers and carbon adsorbers.
         c. open-top vapor degreasers with an open area of greater than 2.0 m^2 should be exempt from installing refrigerated chillers and carbon adsorbers.
       2. Halogenated hydrocarbons are commonly used in vapor degreasing operations because
         a. they are non-flammable.
         b. they are safe in all respects.
         c. they volatilize easily and are easily condensed.
         d. they are far denser than air and are non-flammable.

7.7.   Chemical   Plant/Domestic   Applications-7 (continued).

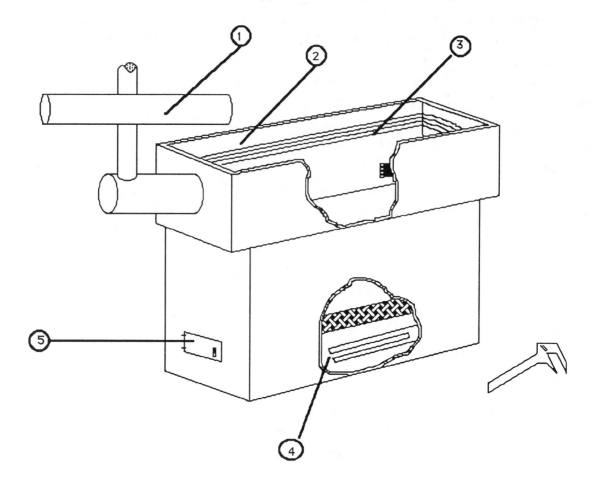

Figure 1.   Schematic of an open-top vapor degreaser.

3. In an open-top vapor degreaser, cleaning is a result of the condensation of _____ on the cool surface of the metal to be cleaned.
   a. dirt, grease and oil
   b. water vapors
   c. solvent vapors
   d. chilling agent

4.   The cooling agent in the cooling coils of an open-top degreaser is usually _____.
   a. liquid nitrogen
   b. freon
   c. methylene chloride
   d. water

7.7. *Chemical Plant/Domestic Applications-7 (continued).*

  5. The freeboard of a vapor degreaser is_____.
     a. also called the lip exhaust.
     b. found above the cooling coils.
     c. used to prevent "drag-out" losses.
     d. all of the above.

d. True/False

  1. Trichloroethylene has excellent solvent properties for cleaning metal surfaces.
  2. A field inspector should never put his/her hand below the vapor level inside a vapor degreaser.
  3. In an open-top vapor degreaser, cleaning action begins when the metal parts reach the temperature of the solvent vapors.
  4. Benzene and carbon tetrachloride are two highly recommended solvents for use in an open-top vapor degreaser.
  5. To reduce "drag-out" emissions, parts should be allowed to dry for at least 15 seconds in the freeboard area.

References:
U.S. EPA (1992), *APTI Correspondence Course 416: Inspection Procedures for Organic Solvent Metal Cleaning Operations*, EPA 450/2-82-014, June.
U.S. EPA (1979), *Solvent Metal Cleaning: Inspection-Source Test Manual*, EPA 340/1-79-008, Chapter 3.

7.8. *Chemical Plant/Domestic Applications-8* (basic principles, unit conversions, partial pressure). [dlu] The Code of Federal Regulations, 40 CFR Part 60, Subpart xx, gives standards of performance for bulk gasoline terminals. This standard describes the collection and processing of vapor displaced from tank trucks being filled. The emissions to the atmosphere are not to exceed 35 mg of total organic compounds/L of liquid gasoline loaded. The tank transfer connections are to be vapor tight so this means the vapor processing system (unspecified) must meet the 35 mg limit. The tank vapor is air saturated with gasoline.

For this problem, assume that the temperature is 75°F, and that 1 L of vapor will be displaced for each L of liquid gasoline transferred. The molecular weight of gasoline vapors formed at 75°F may be taken as 70 g/gmol. The vapor pressure of liquid gasoline is 6.5 psia at 75°F.

7.8. *Chemical Plant/Domestic Applications-8 (continued).*

a. Calculate the partial pressure of gasoline vapor that is equivalent to the 35 mg/L limit.

b. What total pressure would be required for this emission level to be obtained by condensation at 75°F?

c. What fraction of the gasoline would be recovered by a process that reduces the gasoline concentration in the vapor from saturation at 75°F to 35 mg/L?

7.9. *Chemical Plant/Domestic Applications-9* (conservation of mass; material balances; vapor recovery; degreasing operation; PCE emissions). [rrd]   Based on the problem statement and solution for Problem 1.4. Basic Concepts - 4, where should the effort be placed to reduce emissions at a degreasing plant?   In the installation of a vapor recovery system for PCE emissions that is 50% efficient, or for an upgrade of the spent PCE solvent recovery unit to decrease reject PCE flow to 10% of the spent PCE flow rate?   Solve this problem by answering the following questions.   Assume a total demand for PCE of 4,000 lb/yr.

a. Calculate the current emission rate of PCE from this system based on results from Problem 1.4. Basic Concepts - 4.

b. Calculate the PCE emission rate with a 90% efficient solvent recovery system.

c. Calculate the PCE emission rate with a 25% efficient vapor recovery unit and a 75% efficient solvent recovery unit using the material flowsheet shown below.

d. Compare your results and make a recommendation to management as to what system should be implemented to reduce PCE loss from this system.

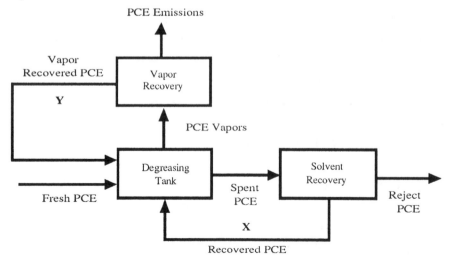

7.10. *Chemical Plant/Domestic Applications-10* (mass transfer, design, solvent losses). [nb] Solvent losses through diffusion of vapor into the atmosphere is a major concern in the vapor degreasing of metal parts (approximately 2 gal/ton of material degreased). Vapor losses are proportional to the temperature of the vapor-freeboard boundary shown in Figure 1. In a simple batch degreasing unit (Figure 1), the primary control of vapor loss in this zone is via a condenser. Fluid flow in the condenser controls the vapor level in the *vapor zone*. Vapor losses may also be controlled by adjusting the height of the freeboard (the distance from the top of the vapor level to the top of the degreaser tank). This problem examines the effect of operating temperature on vapor loss from a degreasing unit.

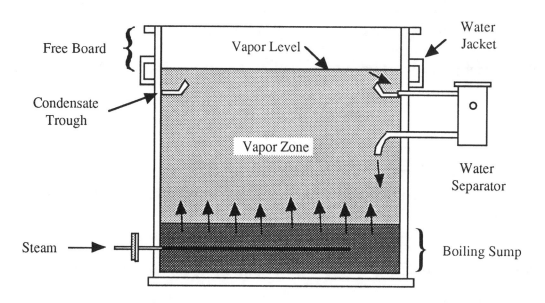

Figure 1. Schematic of a batch degreasing unit utilized for solvent stripping for small parts cleaning. Soil metal parts are suspended in the vapor zone. The vapor condenses onto the parts, dissolving oils/greases. The spent (condensed) solvent containing the grease drains into the hot liquid reservoir. The part is held in the vapor phase until it reaches the vapor temperature, at which point condensation on the part stops. The part dries immediately upon removal from the tank.

a. The built-in condenser in an open-top degreasing tank minimizes losses by keeping the vapor-freeboard boundary temperature at one-half of the boiling point of the solvent trichloroethylene (TCE) in °C. What should the vapor-freeboard boundary temperature be in order to reduce the vapor losses by 50%?

7.10. *Chemical Plant/Domestic Applications-10 (continued).*

Equations 1 and 2 describe the relationship between vapor loss, solvent vapor pressure and temperature. Assume the diffusion of vapor S into air is constant over the pressure range of this problem.

The mass flux of vapor, S, into air, A, can be described as follows:

$$N_{SA} = \frac{D_{SA} P}{RT(h_2 - h_1)} \ln\frac{p_2}{p_1} \tag{1}$$

where D_{SA} = diffusion of S into A, cm^2/s; P = total pressure, mm Hg; p_1 = solvent vapor pressure, mm Hg, at interface; p_2 = vapor pressure at top of freeboard, mm Hg; T = average temperature, K, of the surrounding air; and, $h_2 - h_1$ = height of the freeboard, cm.

The Antoine vapor pressure equation is given as:

$$\ln(p_1) = A - \frac{B}{(T_1 + C)} \tag{2}$$

where A, B and C are the vapor pressure coefficients presented in Table 1 for trichloroethylene and trichloroethane, and T_1 is the temperature at the interface in K.

Table 1. Vapor pressure coefficients for chemicals of interest

Solvent	A	B	C	B.P. (°C)
Trichloroethylene	16.2	3028	-43.2	87.2
Trichloroethane	16.0	3110	-56.2	113.7

b. Repeat Part a for trichloroethane (TCA). If the two solvents are equivalent in cost and cleaning efficiency, which solvent would you select? Why?
c. For this degreasing system, the temperature at the vapor-freeboard interface may be lowered to 25% of the boiling point in °C of the solvent without a significant drop in efficiency. Calculate the maximum possible reduction in vapor loss for TCE.
d. Emission control tests show that solvent losses decrease as the freeboard height increases. Describe the role of freeboard height as a variable in emission control.

7.11. *Chemical Plant/Domestic Applications-11* (separation, membranes). [nb]   Industries using contact process water generate spent aqueous waste streams that often contain hazardous pollutants.   Efficient "detoxification" or removal of the pollutants from these aqueous streams to permit their reuse has become a necessity.   The use of reverse osmosis membrane technology for such separations is a viable option.

A chemical plant generates 1000 L/d of wastewater containing a single organic contaminant.   This organic contaminant, "AOH," is present at a concentration of 0.5 g/L.   The plant plans to concentrate the contaminant in the wastewater to 20 g/L before shipping it for organic recovery.   The wastewater (raw feed) flows into the feed tank and is pumped to several reverse osmosis (RO) membrane units in parallel.   Figure 1 shows a schematic of one of these membrane units.

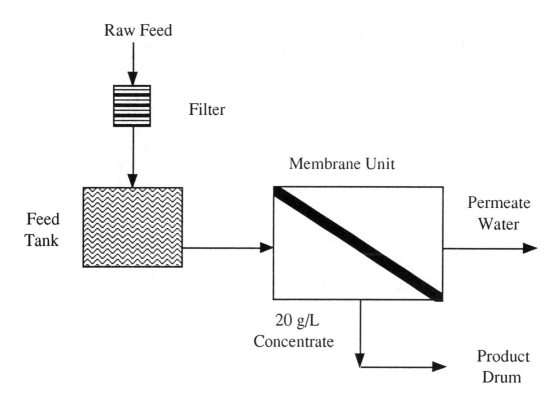

Figure 1.   Membrane separation process schematic.

The contaminant is rejected by the RO membrane and eventually flows into a product drum. The permeate is reused in the plant as contact process water.

A laboratory test was performed on two different polymers, Membrane 1 and Membrane 2.   The results of the organic contaminant flux versus

7.11. *Chemical Plant/Domestic Applications-11 (continued).*

operating pressure, and water flux versus operating pressure are shown in Figure 2.

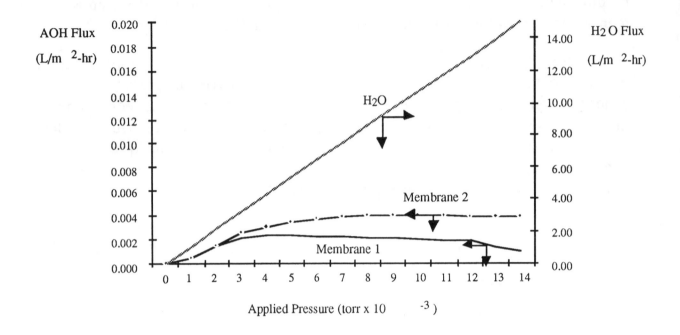

Figure 2.   Separation performance of two thin film composite membranes.

In the separation system shown above, the pressure difference across the membrane is 500 torr and the effective membrane area is 1 m².

a. Using the information in Figures 1 and 2 select the appropriate membrane for the system, i.e., the one with the best contaminant rejection. Determine the permeate flowrate and the concentration of the organic contaminant, AOH, in the permeate. Assume a specific gravity of 1.00 for AOH.

b. Calculate the number of membrane units required to concentrate the AOH from 0.5 g/L to 20 g/L at the given flowrate. For this system assume that the performance of the membrane is not affected by the concentration of AOH in the feed tank.

c. The pK_a of AOH is 10.1, where $pKa = [H^+] [AO^-]/[AOH]$. Suggest an operating pH range for the system. (Hint: For best rejection of the organic contaminant, maximize $[AO^-]$).

7.12. *Chemical Plant/Domestic Applications-12* (recycle, reuse, economics, used tires, vacuum). [ihf, cr] It is important to establish design criteria to aid in the selection of an optimal solution among various processing alternatives. Most often the criterion takes the form of the question: Which of the alternatives would be expected to increase the financial health of the firm by the greatest amount in the reasonably near future? The answer to the question requires an understanding of basic economic principles.

Two commonly used terms are:

Rate of return on investment = Yearly income/total investment

Payback period = Total investment/yearly income

Vacuum pyrolysis of used tires is a new technology being developed at Laval University, Canada. In this process the waste tires are reused to produce valuable products. The low pressure of the process (about 0.15 atm) and short residence time (about 20 s) ensures good quality products, which are estimated as follows:

Product	yield %	lb produced/tire
Crude Oil	55	11 (1.5 gal)
Carbon black	25	5
Steel	9	1.8
Fiber	5	1
Gas	6	1.2

The gas generated is used as the energy source for the process.

The following cost data are based on a vacuum pyrolysis unit for the treatment of 4 tons (8800 lb/hr) of used tires. Each passenger tire weighs approximately 20 lb. Thus, 1 ton of passenger tires is approximately equivalent to 110 tires. As shown above, the products include good quality oil and carbon black. Check if this would be a profitable project by calculating the rate of return on investment, and the payback period using the following data, assuming all costs are in today's dollars.

Assume straight-line depreciation, i.e., withhold from the project earnings equal yearly payments over the projected life of the investment in a non-interest bearing fund sufficient to recover the initial investment.

7.12. *Chemical Plant/Domestic Applications-12 (continued).*

Sales income	$4,700,000/yr
Manufacturing costs	$1,400,000/yr
Total initial investment	$7,000,000
Projected economic life of project	6 yr

7.13. *Chemical Plant/Domestic Applications-13* (recycle, reuse, economics, vacuum pyrolysis, scrap tires). [cr] Vacuum pyrolysis is a new technology to reuse scrap tires and produce valuable products. The low pressure of the process (about 0.15 atm) and short residence time (about 20 s) ensures good quality products, described in Problem Statement 7.12 Chemical Plant/Domestic Applications-12, above. The gas is used as the energy source for the process.

A new plant is to process 4 tons of tire/hr (1 ton = 2200 lb = 110 passenger tires, 1 tire = 20 lb). The following cost data are available:

Investment data for new plant

Fixed Investment	$6,000,000
Auxiliary Investment	$500,000
Working Capital	$500,000

Operating cost data: $/yr = $1,000,000/yr

Business environment

The economic life of the process is estimated to be 6 years. The value of the plant at the end of the 6-year period, its salvage value, is zero. The project is to be part of the business activities of a typical chemicals manufacturer. The plant operates 24 h/d, 260 d/yr.

It is likely that the disposal company charges a tipping (dumping) fee for the used tires, typically about $1 per tire. On the basis of the above cost data, estimate the cost of the crude oil product in cents per gallon, when no allowance is made for risk, profit, or tipping fees. Carry out the calculations for two cases assuming that: 1. carbon black revenue = $0/ton, and 2. carbon black revenue = $350/ton.

For further details on the vacuum pyrolysis of scrap tires see the following article:

7.13. *Chemical Plant/Domestic Applications-13 (continued).*

Christian Roy, Blaise Labercque, and Bruno de Caumia (1990), "Recycling of Scrap Tires to Oil and Carbon Black by Vacuum Pyrolysis," *Resources, Conservation and Recycling*, **4**:203-213.

To estimate the economic value of a proposed system a number of terms needs to be defined.

THE TOTAL INVESTMENT, I. The total investment, I, can be broken down, depending on the degree of risk, into three parts: the fixed investment in the process area, IF; the investment in auxiliary services, IA; and the investment in working capital, IW.

$$I \text{ (in dollars)} = IF + IA + IW$$

These are fixed amounts of money that will be tied up and risked in the interest of the proposed system.

FIXED INVESTMENT, IF. This is the investment in all processing equipment within the processing area. Investment in equipment within the process area carries the greatest degree of risk, since this capital can only be partially recovered as salvage value in case the system is suddenly terminated, for example, when market changes make further operation unprofitable.

AUXILIARY INVESTMENT, IA. Items such as steam generators, fuel stations, and fire protection facilities are commonly stationed outside the process area and serve the system under consideration as well as other processing systems within a large complex.

WORKING CAPITAL, IW. This is the capital tied up in the interests of the system in the form of ready cash to meet operating expenses, real estate, inventories of raw materials and products.

OPERATING COSTS, C ($/yr). These are the costs incurred in keeping the system running from day to day.

PROFIT RATES. The Gross Profit Rate, R ($/yr). The difference between the net income from the annual sales, S (sales receipts less distribution, sales, and promotion costs), and the annual operating costs, C, i.e.:

$$R = S - C \quad (\$/yr)$$

7.13. *Chemical Plant/Domestic Applications-13 (continued).*

The Net Profit Rate, P $/year, defined as the annual return on the investment after deducting depreciation and taxes, is:

$$P = R - e\ IF - (R - d\ IF)\ t$$

where d = a yearly fractional loss of value of the fixed equipment allowed by tax authorities for computing taxable income; e = a yearly fractional assessment calculated to recover the investment in fixed equipment (depreciation); IF = fixed investment, $; R = gross profit rate, $/yr; and t = income tax rate, $/$ earned.

For straight line depreciation e = 1/n, where n = expected project life. The government tax rate t is on the order of 40% of adjusted profit. In the absence of data, d is taken as equal to e.

The experienced rate of return on investment for several processing industries is given below.

Industry	Experienced Rate of Return ($/yr-$)
Pulp and paper, Rubber	0.08-0.10
Synthetic fiber, Chemical and Petroleum	0.11-0.13
Drugs and Pharmaceutical	0.16-0.18

8. Case Studies

8.1. *Case Studies-1* (chemical laboratory, acid waste treatment, neutralization). [hb]   When a small technical college in the northeast constructed a new chemistry and engineering building, the architects included an acid waste treatment facility to neutralize any acid put down the laboratory drains before it was discharged into the sewers.   The regulations require that the waste water have a pH of 2.0 or more before being discharged into the sewer.

The acid waste treatment facility was a small room built around a 200 gal tank which received all water from all laboratory drains.   The tank contained 2500 lb of limestone rocks which occupied approximately 100 gallons of the 200 gal tank.   Thus, when receiving water, the tank contained approximately 100 gal of liquid.   Water from the drains entered the tank at the bottom, rose through the limestone rocks (which reacted with and neutralized the acid) and exited through a pipe at the top in which a pH meter probe monitored the pH.   A pH reading of 2.0 or less triggered an alarm.

The alarm sounded frequently.   Puzzled by the inability of the system to handle the very modest amounts of acid poured down the laboratory drains, the faculty performed some experiments.   When 1.0 molar HCl   (pH = 0) was added to a sample of the limestone rocks, more than 90 min elapsed before the pH rose above 2.0.   Next, a dye was poured down the drain.   Most of the dye exited the tank within six minutes and all of it was gone after 10 min.   Conclusion: the neutralization reaction was too slow to neutralize any significant portion of the acid poured down the drain.

a. Approximately what volume of concentrated hydrochloric acid, HCl, (12 molar) poured down the drain would cause the pH of the water exiting the tank to have a pH of 2.0?

b. Approximately what volume of concentrated H_2SO_4 (18 molar) would have the same effect?

c. The college lacked the resources to replace the acid neutralization facility.   What could it do to comply with the requirement of not discharging water with a pH of 2.0 or less into the sewer?   Shutting down the laboratories or eliminating all experiments that use acids is not an acceptable solution.

8.2. *Case Studies-2* (tannery waste, filter press, sludge dewatering, chromium sludge). [kg] A tannery uses a filter press to dewater its raw sludge. The dewatered sludge contains 120 mg of total chromium/kg of sludge. The filter press has a surface area of 150 ft^2 with a filtration rate of 10 gal/hr-ft^2. Assume that it is 100% efficient in separating solids. Estimate the amount of total chromium disposed/yr from this plant if the plant operates 16 hr/d, and 7 d/wk, 50 wk/yr. The following information is available to you.

Before dewatering:
Water content of sludge = 95%; Solid content of sludge = 5%

After Dewatering:
Water content = 52%; Solid content = 48%

8.3. *Case Studies-3* (design, kinetics, petroleum refining, hydrodesulfurization, sour crude, hydrogen sulfide). [nb]   Gradual depletion of preferred petroleum crudes has spurred an interest in the evaluation of processes for the difficult-to-treat sour crudes. The use of heavy distillates from sour crude for the production of petroleum is complicated by the presence of sulfur.   Many refineries have developed desulfurization systems to process heavy distillates obtained from the less expensive sour crude. One such system, "Catalytic Desulfurization" (Figure 1), is designed to meet current limits on stack emissions of sulfur and sulfur-products. Typically the feed to system is 5 to 15 wt% S.

a. For a typical sour crude distillation column a simple three output system is shown in Figure 2. The preheated crude has a specific gravity, s = 0.958, and is flashed into a column at the rate of 24,000 bbl/d. The three output streams are: (i) lights (gas phase), density = 0.185 lb/ft^3; (ii) gas/oil phase, s = 0.810, and; (iii) heavy distillates, s = 1.15. Using the information in Figure 2, calculate the mass and volume flowrate of the heavy distillates. Note: 1 bbl = 42 gal.
b. Determine the energy, required to heat the heavy distillate from 500°F to the hydrodesulfurization temperature of 950°F. The estimated heat capacity of the heavy distillate fraction is given by: C_p = s (45.2 + 9 x 10^{-2} T) BTU/bbl, where s = specific gravity of the heavy distillate fraction, and T = °R.

8.3. *Case Studies-3 (continued).*

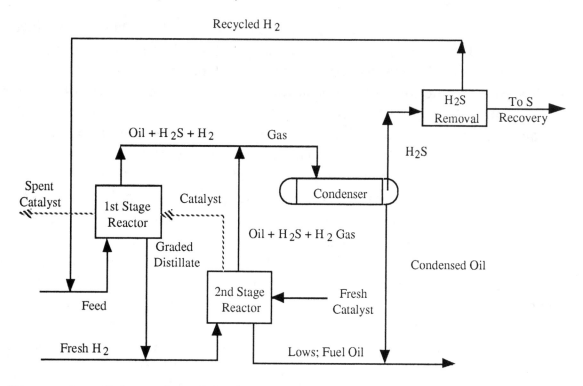

Figure 1.   Schematic of hydrodesulfurization unit for petroleum desulfurization.

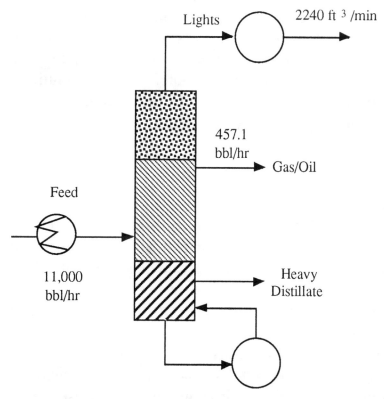

Figure 2.   Schematic of unit for distillation of sour crude.

8.3. *Case Studies-3 (continued).*

    c. Calculate the energy cost if natural gas is used to generate the heat in Part b. The heat value of natural gas is 940 Btu/ft^3, the heating efficiency is 0.8, and the cost is $1.50/MSCF.

    d. Design a single reactor to desulfurize 450 bbl/h of heavy distillate from 10 wt% S to 0.50 wt% S. Assume: first order kinetics in the catalytic hydrodesulfurization system with a rate constant k = 0.40/hr; and 20 vol% bed expansion for the reactor.

    e. An alternative to the hydrodesulfurization of the crude oil is to install pollution control devices on the combustion exhaust stacks. The total long term cost of pollution control devices is estimated to be $5.00/ton of SO$_2$ gas trapped. The annualized capital cost (the annual charge to service debt on the total capital investment) for the reactor designed in Part b is estimated to be $0.235/gal of reactor volume. The annual cost of catalyst, regeneration of the catalyst, hydrogen and operation (excluding energy) of the reactor is 30% of the annualized capital cost. Based on these figures, which process is preferred? Why?

8.4. *Case Studies-4* (circuit board manufacturing, hazardous waste, pollution prevention, rinse water, economic feasibility, design). [kg] A printed circuit board manufacturer has hired you as a consultant to evaluate pollution prevention measures in their plant. With the following data available to you, evaluate the potential for pollution prevention in this plant, especially for the rinse water. Perform a simple cost analysis to show economic feasibility of the recommended pollution prevention measures.

Plant Description:

The printed circuit board manufacturing process includes drilling and routing, layering, photoresist printing, plating, etching and stripping. Plating, etching and stripping processes generate hazardous waste. The maximum amount of aqueous waste is generated from the rinse tanks. The treated sludge from this wastewater is classified as hazardous waste and is containerized and disposed off-site. The plant uses nine dip rinse tanks and three spray rinse tanks. An on-site deionization plant supplies the rinse water. The flow rate of water through each dip rinse tank is 16 gpm. The flow rate of water through the spray rinse tank has been estimated to be 1.5 gpm.

8.4. *Case Studies-4 (continued)*.

The wastewater is treated in an on-site industrial waste facility. Chemical treatment is performed in three separate tanks. In the first tank ferrous sulfate is used to reduce the copper to its precipitable form. Alum and sodium hydroxide are added in the next tank. Alum addition causes the suspended solids to coagulate, forming larger particles. In the third tank a polyelectrolyte coagulant is introduced to aid in the flocculation of the contaminants into large flakes for easy settling and separation. The sludge is dewatered and disposed of, and the wastewater is discharged to a POTW. Sludge is dewatered in a bag filter that increases the solids content to 11%. The analysis of the dewatered sludge gave the following results:

Percent solids - 11%

Metal	%
Copper	9.75
Nickel	0.30
Tin	2.30
Iron	19.95
Zinc	0.45
Lead	1.15
Chromium	1.00
Total Metals	34.90
Other compounds	65.10

Chemical usage in the waste water treatment plant includes the following:

Ferrous sulfate:	85 lb/1000 gal of waste @ $0.33/lb
Alum:	85 lb/1000 gal of waste @ $0.47/lb
Sodium Hydroxide:	20 gal/1000 gal of waste @ $0.74/gal
Polyelectrolyte:	0.3 lb/1000 gal of waste @ $7.50/lb

The plant presently operates its dip rinse tanks as flow-through tanks. Deionized water is plumbed into the tank during operation and the overflow is plumbed to the collection sump. The general industry standard suggests that the concentration of chemicals in the rinse solution must not exceed 1/1000 of the concentration of the chemicals in the process bath. The drag-out rate of chemicals in a printed circuit board manufacturing process is approximately 15 mL/ft^2 of board. An average work piece rack

8.4. *Case Studies-4 (continued)*

holds 2.5 ft² of board, and the printed circuit board is kept for 3 min in the rinse water.

8.5. *Case Studies-5* (solar cooling, general principles, thermodynamics, refrigeration). [glh] As an alternative energy source, the sun has received more attention in recent decades because of its non-polluting character, wide availability and low cost. Solar radiation has been used to generate electricity, either directly or through the vaporization of water to run steam generators. It can be used in domestic and commercial hot water heating for space heating and for washing. It is also possible to use solar energy to provide cooling in the air conditioning systems of buildings. One such project was funded by the NSF in 1974 for use in the Dade County School in Dade County, Florida (Martinez, 1981).

Whenever a liquid has a solid dissolved in it, its vapor pressure decreases according to Raoult's Law. This principle can be used to generated a region of low vapor pressure in a closed loop system. The system illustrated in Figure 1 uses the low vapor pressure of water over a concentrated aqueous salt solution (lithium bromide) to cause the evaporation of pure water in an attached vessel. The evaporation of the pure water provides the cooling source for an air conditioning system. Four major sections comprise the total system:

1. The evaporator, A, where heat is absorbed from the hot school by the evaporation of water at this reduced pressure;
2. The absorber, B, in which the concentrated LiBr solution is sprayed through the incoming water vapor, absorbing it and generating heat through condensation and dilution;
3. The generator, C, which receives the more dilute salt stream, treats it with the energy (steam) from the hot water holding tank, and regenerates the concentrated salt solution by evaporation; and
4. The condenser, D, which is simply a condenser for the vapor produced in the generator, returning it to the evaporator, A.

The cyclic process is more simply illustrated in Figure 2.

8.5. *Case Studies-5 (continued)*

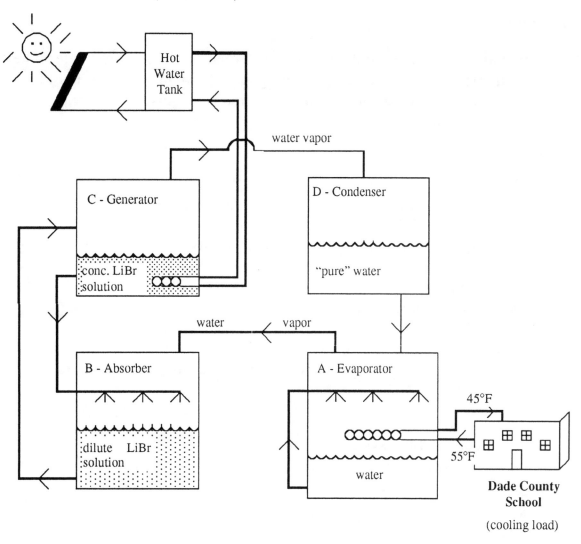

Figure 1. Detailed schematic of solar cooling system utilized in the Dade
County School, Dade County, Florida.

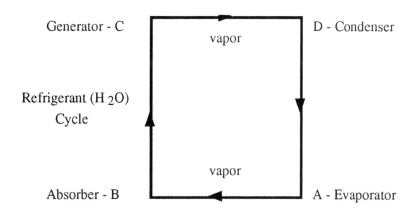

Figure 2. Simplified schematic of refrigerant water cycle in solar cooling
system.

8.5. *Case Studies-5 (continued)*

Determine the following:

a. Calculate the energy output from the solar collector and determine if it will provide the necessary cooling power to meet the school's requirements. At what minimum efficiency must the refrigeration system be run?

Data:

Solar collectors: Area = 1730 m², efficiency = 91%

Solar radiation, watts/m²: July (max) 260, Dec. (min) 154, Average = 219

Cooling load demands (Ave. daily load, Btu/hr): Sept. 652; Dec. 611; June 560

3.413 Btu/hr = 1 watt

Assume a 30-day month and a 12-hr day.

b. Calculate the pump speed in gal/min required in order to maintain the 45° refrigerator output to the school during the month of September.

Data:

The school cooling system contains 150 gal water

Heat capacity of water = 1 cal/g-C°

1 gal = 4 qts = 3.78 L

1 g of water = 1 cm³

9/5 C° = (°F - 32) or ΔC° = 5/9 ΔF°

1 cal = 4.184 J

1 watt = 3,600 J/hr

c. Determine whether the LiBr/H₂O refrigeration system is economical by calculating the payback time. Usually, a payback period of 5 to 10 years is considered economical.

Data:

	LiBr/H₂O	Conventional A/C
Capital Costs:	$250,000	$200,000
Installation Cost:	35,000	20,000
O&M Cost/yr:	850	375 plus*

Cost of electrical power: $0.15/kW (assume that it is constant.)

*Estimated annual power consumption of conventional system: 36,500 kW/yr.

8.5. *Case Studies-5 (continued)*

For the purpose of this problem, also assume that no money is borrowed (or that money is free.)

d. A reduced pressure of no more than 0.27 mm Hg is required from the concentrated solution of LiBr for this system to work as designed. This is approximately a 55% solution (by weight). Calculate the molality and the mole fraction of the salt solution. Given the relationship that the vapor pressure, P, over an aqueous salt solution is equal to the pressure of the pure solvent at the given temperature, $P^{\circ}_{solvent}$, times its mole fraction, $X_{solvent}$, (i.e., $P = P^{\circ}_{solvent} X_{solvent}$) calculate the theoretical pressure over the LiBr solution. Assume ideal behavior.

Data:

The temperature of the concentrated LiBr solution is 115°F

T, °C	Vapor Pressure for pure water, torr
32.0	35.7
46.0	75.6
60.0	149
82.0	385

1 mm Hg = 1 torr

References:

Martinez, A.R., Ed. (1981), *Solar Cooling and Dehumidifying*, Pergamon Press, New York.

Jennings, B. H. (1978), *The Thermal Environment-Conditioning and Control*, Harper and Rowe, New York.

8.6. *Case Studies-6* (vapor recovery, gasoline station, fugitive emissions, cost analysis). [dlu] A gasoline service station owner has a station in a nonattainment area as defined by the U.S. EPA. This means that ozone concentrations are high and equipment will soon have to be installed to recover the gasoline vapors displaced from the automobile tanks during filling. Assume that a typical station pumps 3,000 gal of gasoline/d and operates 6 d/week for 52 weeks/yr.

8.6. *Case Studies-6 (continued)*

The gasoline vapor collected from displacement during filling could be recovered by compression and condensation. Scavenger hoses will need to be installed to pass the vapors from the automobile tank back to the supply tank at the same time as the gasoline flows from the supply tank to the automobile tank. The vapors are then recovered from the supply tank each evening as the supply tank is refilled. Filling the supply tank takes about 30 min. The procedure is to compress the air–gasoline mixture as the supply tank is filled until the dew point is high enough to obtain condensation with the cooling medium available. Assume for simplicity that the compression can be done isothermally at 60% efficiency so that the compression power, P, is:

$$P = C \, W \, R \, T \, \ln(P_2/P_1)/0.60$$

where C is a conversion factor, W is the molar flow rate, lbmol/sec; R is the gas constant, 10.73 psi-ft^3/lbmol-R; T is the temperature, °R; and P is the pressure, psia.

Ninety percent of the gasoline vapor is to be recovered by condensation at 75°F. The primary cost of operating this unit will be the $0.10/kW-hr cost of electricity to drive the compressor. The cooling water cost may be neglected at this level of analysis.

The total capital investment may be estimated as the annual cost of investment divided by the capital recovery factor. The capital recovery factor, CRF = $i(1+i)^n/((1+i)^n - 1)$ with an interest rate i = 0.1 and a useful life of n = 10 yr.

Assume the automobile tank vapor space, the air and the gasoline supply are all at 75°F. The vapor space is saturated with gasoline. The vapor pressure of gasoline under these conditions is about 6.5 psia. The vapor has a molecular weight of 70 g/gmol when condensed, and has a liquid specific gravity of 0.62.

The station owner is willing to pay $1.00/gallon for the recovered gasoline. How much can the station owner afford to invest in capital equipment if the sum of the operating and annual investment costs is to remain less than $1.00/gal? Do you think the station owner can make a profit from this recovery operation? Explain. BP Oil estimates that it will cost $25,000 in 1993 dollars to retrofit a station for vapor recovery.

8.7. *Case Studies-7* (exposure, TLV, TCE, VOCs, dry cleaning operations). [smk] In a dry cleaning operation, an employee uses trichloroethylene (TCE) to remove stains from clothing. TCE has been identified as a carcinogen. Also, since it is a volatile organic compound (VOC), emissions of TCE to the atmosphere contribute to photochemical smog formation. EPA has placed TCE on its toxics lists and this material is targeted for a 50% reduction in emissions by 1995.

Estimate allowable worker exposure and emissions rates from dry cleaning operations by answering the following questions. The information below is needed for this estimation.

Vapor pressure of TCE (from CRC Handbook): log P (torr) = (−1816.8/T) + 7.95642, where T is in Kelvin;
molecular weight of TCE = 131.39 g/gmol;
TLV data: TLV TWA = 50 ppm, TLV STEL = 200 ppm

*NOTE: The "TLV" is the threshold limit value or the exposure concentration below which there are not adverse health effects to the exposed individual; "TWA" is the maximum allowable, 8-hr time-weighted average exposure concentration; "STEL" is the short-term exposure limit defined as the 15-min TWA exposure which should not be exceeded at any time during an 8-hr work day.

The time variation of the concentration of a species emitted into a volume V may be calculated from:

$$\frac{dC}{dt} = \frac{G(t)}{V} - kQ\left(\frac{C}{V}\right)$$

where C = concentration of species at time t, kg/m^3; G(t) = emission rate of species at time t, kg/s; V = volume of room, m^3; Q = room ventilation rate, m^3/s; and k = mixing constant for imperfect mixing, with k=0.3 at height for stagnant conditions. This equation assumes that the concentration of the species is 0 in the ventilation air.

a. Suppose no ventilation of the workroom occurs. Assuming a constant usage rate, how much TCE can evaporate during a typical 8–hr workday before the recommended maximum exposure is exceeded? Assume the room is 4 m wide, 5 m long, and 4 m in height.

8.7. *Case Studies-7 (continued)*

b. Assume that actual evaporation rates of TCE average 50 g/d. What minimum ventilation rates are needed to maintain the steady–state room concentration of TCE below acceptable values?

c. Assume that, on the average, there are two dry cleaning stores per urban mi^2, and that the stores operate 6 d/wk. For the ventilation rates computed in Part b, what annual emission rates can be attributed to dry cleaning in a major metropolitan area, which has an area of approximately 100 mi^2?

d. Since the evaporation rate is proportional to vapor pressure, the emissions might be controlled by lowering the temperature at which the TCE is held. What reduction in temperature (from the ambient temperature of 25°C) would be required to halve the emissions rate? What temperature is required to reduce the vapor pressure to 10% of that at 25°C?

e. As part of a pollution prevention strategy, it is suggested that the TCE vapors might be recovered from the ventilation stream by chilling at constant (atmospheric) pressure. Evaluate the feasibility of this suggestion by computing the temperature to which the stream must be cooled. Assume the room concentration of TCE is equal to 25% of the TWA value.

8.8.   Case Studies-8 (waste treatment investment, optimization, cost analysis). [ap] In most instances, companies are faced with a choice between paying for disposal of hazardous wastes they generate, or to purchase equipment to render these wastes non-hazardous so they are more easily and less expensively disposed of. Treatment options often involve recovery and recycle of a component (e.g., solvent) of the process, and can result in substantial savings in raw materials costs. An additional consideration is the reduction in corporate expenses for insurance, fines and legal support that is often a result of the investment in pollution prevention equipment.

As part of the waste treatment/pollution prevention decision-making process, the economic implications of purchasing this waste treatment/recycling equipment must be considered. Assume that, for a particular case, the relation, $p\,l^4 = k$, holds, where p = the initial investment cost for equipment, l = the annual liability cost (cost of violations of environmental regulations) minus annual operating and maintenance costs, and k = a constant.

8.8. *Case Studies-8 (continued)*

a. What is the optimal investment in equipment for a 20-year time period at an interest rate of 5%? NOTE: The present value of the investment should be optimized, subject to the constraint that the p and l vary inversely according to the relation given above. The optimal value of p should be expressed in terms of the constant, k.

b. Repeat the calculation for interest rates of 10%, 15%, 20%, and 25%, and present the results in graphical form.

c. Repeat the calculation for all interest rates, assuming a 5-year investment period, and plot these results on the same graph as that generated in Part b. How does the assumption of a shorter investment period affect the optimal initial equipment investment?

9. Ethics

9.1. *Ethics-1* (ethics, definitions). [ihf] Most pollution prevention regulation depends on voluntary compliance. Hence, ethics becomes a pollution prevention concern. In dealing with ethics there are several terms that one may encounter: ethical theory, consequentialist theory, deontological theory, utilitarianism, ethical egoism, nationalism, and retributivism. Briefly explain these terms.

Reference: W. K. Frankena and J. T. Gremrose (1974), *Introductory Readings in Ethics*, Prentice Hall, Englewood Cliffs, New Jersey.

9.2. *Ethics-2* (ethics, existential engineering). [ihf] In the 1960s there was an anti-technology movement in which engineers were blamed for the ills of our society. Engineers were blamed for nuclear bombs, pesticides, crashes, etc. This is sometimes described as the "Existential Pleasure of Engineering." Explain this description, and discuss how it pertains to pollution prevention.

9.3. *Ethics-3* (toxic chemicals, regulatory matters, ethics). [hb] A chemical company's internal environmental audit uncovered a significant pollution problem involving repeated releases of a toxic chemical into the sewer system. The company took immediate action and completely corrected the problem but did not inform any regulatory agency. Discuss the ethics of the company's actions.

9.4. *Ethics-4* (governmental liability, personal liability). [hb] The federal government considers its employees to be <u>personally</u> responsible for violations of environmental laws. The government reasons that since it is not possible for an employee to be required in his/her job description to violate any law, rule or regulation, then any such violations must have been committed while the employee was acting "outside the scope of his or her employment." If the employee acted outside the scope of his or her employment, then he/she is not entitled to legal help from the government and must personally pay any fines levied. If the employee is sent to jail, he/she would almost certainly be fired. Managers who directed or allowed employees to violate environmental laws, rules, or regulations would be held even more responsible than the employees that he or she supervised.

9.4. *Ethics-4 (continued)*.

Claims by managers that they didn't know what the employee was doing have routinely fallen upon deaf ears.

This policy constitutes a punitive employee incentive program. Give at least one benefit of this policy and at least one drawback of this policy. Suggest some ways that the incentive program could be improved.

9.5. *Ethics-5* (environmental ethics, code of ethics, professional ethics). [dem] Ethical situations may involve some of the hardest decisions we as engineers and scientists will ever have to make. The American Society of Civil Engineers describes, in part, the Fundamental Principles of their Code of Ethics (effective January 1, 1977) to be the following:

"Engineers uphold and advance the integrity, honor, and dignity of the engineering profession by:
1. Using their knowledge and skill for the enhancement of human welfare;
2. Being honest and impartial and serving with fidelity the public, their employers, and clients;
3. Striving to increase the competence and prestige of the engineering profession; and
4. Supporting the professional and technical societies of their disciplines."

But how do we as engineers and scientists make the decisions on a case-by-case basis to be "ethical"? Using the following information, describe how to ethically handle each situation.

OR - Divide into groups giving each person in the group one of the following roles and have the group(s) arrive at a solution that can be defended to the other groups.

ROLES:
Client; engineer/scientist/friend; regulator; lawyer friend; engineering consultant.

CLIENT: You, as a client, have a small (500 gal) underground storage tank for diesel fuel on your farm. On January 1 you fill the tank for use

9.5. *Ethics-5 (continued).*

in the following months. Historically, you know you use approximately 150 gal a month to run the farm.

On February 15 of the same year, you drive your tractor to the tank to fill it up for the days work. However, there is no fuel to fill the tractor.

What do you do?

You are three years from retirement, you own your house and the farm free and clear. You have $30,000 saved in the bank for your retirement. You know a couple of engineers but aren't sure if you should contact them to help you.

You have a friend who is a lawyer.

Your tank has been in existence for approximately 15 years and is unregistered with the state authorities.

OPTIONS:
* Report the situation to the state as soon as possible.
* Call your engineering friend.
* Call your lawyer friend.
* Remove the tank and surrounding soil without reporting the incident. Spread the soil on the ground and keep the animals away. Cover with plastic to avoid rain water from leaching fuel into the soil.
* Remove the tank and surrounding soil without reporting the incident. Spread the soil on plastic to vent it to the atmosphere and keep animals away. The plastic underneath is to prevent leachate from further contaminating the ground water.

ENGINEER/SCIENTIST/FRIEND: A friend comes to talk to you about a problem he is having with a leaky underground storage tank he has on his farm. You know a little about the process of removing underground tanks, the necessary testing which goes along with their removal, and the state regulations for USTs. You know a consulting engineer who specializes in underground tanks in the state.

9.5. Ethics-5 (continued).

OPTIONS:
* Advise your friend to call the state.
* Call your consulting friend on USTs.
* Advise your friend to the best of your knowledge.
* Advise your friend to call his lawyer.

ENGINEERING CONSULTANT: You know all the procedures and the regulations of the state to properly handle leaking underground storage tanks. You know that there is state funding for the removal of underground tanks but that the funding only exists for corporations/industrial facilities rather than for the individual. You also know that all leaking underground tanks must be reported to the state. You are also working with two other farmers who have reported their tanks to the state and who are now being held responsible for cleaning them up. However, they do not have the financial capability and are therefore working together with a lawyer to attempt to either get funding for individuals from the state or obtain at least enough funding to simply clean up the immediate danger and leave the rest to simply volatilize into the atmosphere through time.

OPTIONS:
* Give advice to your engineering friend without knowing the client.
* Advise them to contact the state immediately.
* Advise them to remove the tank and immediate surrounding soil, and then contact the state
* Tell your supervisor of the phone call.
* Coordinate with the other two farmers and their lawyer to see if they will allow another farmer to join their efforts.

LAWYER; FARMERS FRIEND: You listen intently and give one of the following responses:

* Advise farmer to report to the state immediately.
* Advise farmer to do nothing.
* Advise farmer to remove tank and surrounding soil without reporting it.

REGULATOR: Send farmer all procedures and financial responsibility information knowing there are no financial means of assisting him. Press state regulations requiring proper tank removal and site remediation.

9.6. *Ethics-6* (environmental ethics, NIMBY, Native Americans, reservations, landfill siting). [dem] As the nation quickly runs out of landfill space for waste disposal and the NIMBY (not-in-my-backyard) syndrome generates increasingly strong public opposition to siting of new landfills, prospective developers have turned to communities more likely to accept waste. Native American territories have become especially attractive for "development" proposals for waste management; more than 30 reservations have been targeted for incinerators or landfills in the recent past.

Many reservations in the United States are some of the poorest communities in the world. Unemployment runs at 75% or higher, and alcoholism, drugs, teenage pregnancy, inadequate education, and a host of other social problems inflict great damage to the community as a whole. The special sovereignty status of Indian nations provides a means of bypassing state regulations since tribal governments lack the resources, institutions, and personnel to ensure environmental compliance. In some cases, negotiations may benefit from the nation's internal political divisions between a tribal council with close ties to the Bureau of Indian Affairs, and the traditional leaders. Corruption is also known to be common place in some tribal governments.

Like all indigenous peoples in the world, traditional Native Americans have a special relationship with the land as they place spiritual emphasis on environmental preservation. Overconsumption, environmental destruction, and the ensuing need for massive waste disposal is therefore contrary to those beliefs.

Is it ethical for a landfill developer to try to build a facility on Indian land knowing that less regulatory burden is likely to result from inadequate (or nonexistent) environmental enforcement, corruption, or simply the dire need for some means of subsistence? Should communities that for centuries strove to preserve harmony with the environment, be doomed to become the waste repositories for the industrial society that initiated environmental destruction?

Would it be ethical for an engineer to accept a job to design a facility that does not use pollution prevention methods that would be required outside the reservation but are not required on the reservation?

9.6. *Ethics-6* *(continued)*.

Is it ethical for the United States to allow hazardous waste to be shipped across its border into the sovereign Indian Nations knowing that it will not be properly treated on the other side of the border?

9.7. *Ethics-7* (environmental ethics, international trade, Mexico, international environmental regulations). [dem] As international trade boundaries become blurred by expanding global markets, some U.S. chemical manufacturers have moved operations across the Mexican border in order to lower labor costs and improve competitiveness. An additional attractive feature is the less stringent Mexican environmental regulations.

Despite the existence of these local regulations, some companies have been illegally disposing of hazardous wastes into the ground and surface waters such as the Rio Grande. Despite denials from headquarters in the U.S., unusually high incidences of birth defects such as anencephaly, babies born without a brain, have been reported on both sides of the border and are believed to be linked to water contamination. In spite of health effects on U.S. populations, the jurisdiction of U.S. officials ends at the border. Actions from their Mexican counterparts are, despite the protests from the Mexican population, unlikely since the presence of the companies represents employment and a source of income for the towns.

Are these companies acting unethically?

9.8. *Ethics-8* (toxicology, exposure assessment, risk assessment, health effects). [hb] Many toxicologists doing long term testing of toxic chemicals have observed that the test animals exposed to the lowest dose of the toxic chemical actually had better health than the control group of animals that had no exposure at all. In 1967, Dr. H.F. Smith Jr. proposed an explanation of this effect which he called "sufficient challenge." Briefly, the theory is that small amounts of the toxic chemical challenges the animal's ability to maintain homeostasis - the ability to maintain an internal balance of body chemical processes and an independence of the environment. According to this theory, the animal's response to the challenge of the smallest dose of toxic chemical actually improves its vigor. When, however, the exposure concentration of the toxic chemical rises above the animal's ability to adjust, injury occurs and health suffers. If this theory is correct, might it be used to argue against environmental pollution laws and regulations?

9.8. *Ethics-8 (continued).*

What factors would favor continuance of environmental pollution abatement laws and programs even if Dr. Smith's theory is correct?

9.9. *Ethics-9* (ethics, environmental audits, liability). [wrl] You are Director of Environmental Affairs for Sludge Chemical Co. Your job is to implement the environmental policies established by executive management. As a middle manager, you have no say in setting these policies, but you agree with the policy decisions that have been made. Thanks to these policies and your effective implementation, Sludge Chemical Co. has an exemplary environmental record and has never been cited for non-compliance with a regulation.

The CEO of Sludge says that he wants to continue to assure compliance with all state and federal rules and has announced the formation of a new department, Internal Environmental Audits (IEA). This group will serve as a company-wide watchdog and will report directly to Sludge Chemical's outside legal counsel who will then report to the CEO. Restating what a wonderful job you have done, the CEO now grants you complete dominion over all things environmental. Henceforth, any problems will be reported by IEA via Sludge's attorney, so there is no need for you to call with details.

What reasons can you give for the new department and it's unconventional structure? Are you uncomfortable with this situation? If so, why?

9.10. *Ethics-10* (ethics, recycling, incineration). [wrl] You are employed as director of environmental operations for Biguns Chemical Co., a manufacturer of various commercial chemicals sold as feedstock for other processes. One of the waste products produced by your plant is classified as a RCRA hazardous waste due to its flammability and reactivity. Your staff has just completed a review of the company's options in dealing with this waste material and the cost associated with each, including risk assessments and estimates of future liability. You have asked your staff to present the results as simply as possible for purposes of management discussion. Following are the options:

9.10. *Ethics-10 (continued).*

Option 1. Process modification or product reformulation at an equivalent disposal cost of $1,795/ton. The result will be a non-hazardous waste.

Option 2. Waste solidification and construction of an on-site hazardous waste monofill at a cost of $1,538/ton.

Option 3. Incineration of the waste by a licensed TSD facility at a cost of $1,410/ton.

Option 4. "Recycling" of the waste by a non-licensed facility using a proprietary vitrification process utilizing combustion in a rotary cement kiln at a cost of $1,282/ton. The vitrified material can be sold as an aggregate for use in road construction at a price of $35/ton.

Upon closer inspection, you learn that the recycler in Option 4 has been accused by EPA of being a sham recycler, although a lengthy court battle has recently ended in a stalemate on this issue. If you choose to "recycle" your waste under Option 4, the worst case scenario outlined by your staff is that Biguns' would have to retain possession of the vitrified waste, which is maintained in a discreet stack by the recycler. This material passes all applicable EPA hazardous waste tests.

Assume that action under Options 2, 3 or 4 is at your sole discretion while use of Option 1 would require approval of an investment committee composed of executive management. Which option will you choose? Offer a brief explanation of your reasoning.

10. Term Papers and Projects

10.1. Term Papers & Projects-1 (ethics, environmental racism, minority populations, socio-economic status). [dem] The 1987 *Toxic Wastes and Race* study carried out by the United Church of Christ's Commission for Racial Justice uncovered the following findings (see reference):

- Three out of five African Americans live in communities with abandoned toxic waste sites.
- Fifteen million (60% of this population) African Americans live in communities with at least one toxic dump site.
- Of the the five largest commercial landfills in the country, three of them, accounting for 40% of the nation's 1986 capacity, are located in predominantly African American or Hispanic communities.
- Race is the single most important factor, more than income, social class, or property value, which correlates with the location of toxic waste sites.

Similarly, the Green Peace report *Playing with Fire* indicated that minority populations are 89% higher than the national average in communities where incinerators operate, and 60% higher in communities where incinerators are proposed. Other sources single out migrant farm workers as being exposed to the worst effects from pesticide spraying, inner-city residents as likely to be exposed to higher environmental risks, and Navajo teenagers as having 17 times the national average of organ cancer incidence, most likely as a result of uranium mining.

These facts highlight the issue of "environmental racism." Write a paper on the difficulties experienced by minority communities in confronting environmental contamination in view of the lack of financial resources, political clout, and trained environmental professionals in those communities.

Suggested reference:
Boullard, R. D. (1992), "In our backyards," *EPA Journal*, 18(1), p. 11- 12.

10.2. Term Papers & Projects-2 (solid waste management). [lar] Determine the solid waste disposal practices used in your home town. Write a report describing your findings. The report should include the following:

10.2. *Term Papers & Projects-2 (continued)*

1. A history of collection and disposal.
2. A description of the current collection systems, i.e., curbside, once per week, city operated, private hauler, etc.
3. The per capita generation rate (to determine your own per capita generation rate, weigh the garbage bags you would put on your curb for one week and divide by the number of people in the household).
4. Answers to the following questions: Where does your solid waste go? Does your community use a transfer station? How long is the waste in storage and transfer? Does your community have its own landfill? What is the cost of transportation?
5. If your community has its own landfill, or transports its waste to a facility owned by another entity, answer the following questions about this landfill: What are its current practices related to daily cover and leachate and methane control? Is the landfill lined? With what? Does the facility have a leachate collection system? Does it have a ground water monitoring system? How big is the site? What is its life expectancy? What are the tipping fees collected at the facility, and are they adequate for current needs plus future liability and closure requirements?
6. A discussion of the cost to residents for service.
7. Comments and suggestions for improvements (yours or someone you've talked with to get the information needed for the report).

10.3. *Term Papers & Projects-3* (municipal wastewater treatment plant sludge, sludge management). [lar] Determine the sludge disposal practices of your home town. Write a report describing your findings. The report should include a discussion of the following items:

1. The quantity of sludge produced as well as seasonal variations, if any, experienced at the plant.
2. On-site temporary storage. If the plant does provide for temporary storage describe the capacity, whether it is covered or not, and any management techniques that are utilized to control drainage from the sludge storage areas, used to treat the drainage, used to control odor problems. Also discuss whether the capacity varies with seasons.
3. Discuss sludge disposal options. Where is the sludge disposed of and how? What quantity of sludge is disposed of? What regulations control its disposal? What are the costs of disposal?

10.4. *Term Papers & Projects-4* (recycling, markets). [lar] Many communities are beginning (or have begun) a voluntary or mandatory recycling program. These may be private, industrial, public, or domestic systems. But now that everyone is on the bandwagon of saving "mother earth," how do we keep/develop the markets for these now recycled goods. What is the incentive for a company (public or private) to begin/continue the recycling of these now cleaned, separated and gathered goods. Unfortunately, often our recyclables are picked up at our curb, taken to a transfer station, and there is no market, or rather uneconomical markets, for all the goods we have so carefully cleaned and separated.

Suggestions for beginning the project:
a. See where your recyclables go after they leave your curb. Are they sent to a recycle market place or simply stored temporarily at a transfer station and eventually disposed of in a landfill?
b. Determine if there are any incentives (financial) within your state recycling regulations which assist in starting new recyclable markets or help those that are in economic trouble to maintain their existence.
c. Is there anything we as engineers/scientists and consumers do to assist the development of or continued life of the current recyclable markets?

10.5. *Term Papers & Projects-5* (vacuum pyrolysis, recycling, waste tires, regulations). [cr] Vacuum pyrolysis is a thermal decomposition process which successfully decomposes scrap tires into useful products. The process is conducted in the absence of air and consequently does not involve the combustion of the tires. The pyrolysis products are found on a wt% basis to be composed of the following materials:

Oil, 55%; Carbon black, 25%; Steel, 9%; Fibers, 5%; Process Gas, 6%

The process gas is used as a make-up heat source for the pyrolysis reactor. The oil, carbon black and steel are valuable recyclable products. The fibers can be disposed of as nonhazardous wastes.

A capital venture corporation is willing to invest $10,000,000 to build and operate a demonstration unit. The throughput capacity of the contemplated plant is 3000 kg/h. There are 110 passenger tires/metric T. Two different locations are considered by the investor: New York and

10.5. *Term Papers & Projects-5 (continued)*

California.   Other than the end-market for the products, one area of
concern for this particular investor are the federal and state regulations
that the owner of the plant will have to comply with once the plant is fully
commercial.   In your evaluation of this problem, answer the following
questions: (Note, this is an open-ended question and current regulations
should be reviewed to provide a current and accurate response to the
problem).

   a. What are the regulations under the Clean Air Act governing this
      plant at both sites?
   b. What are the regulations with regard to the ultimate disposal of the
      fiber solid residues in California?... in New York?

Supplementary information about the process can be found in C. Roy et
      al. 1990. "Recycling of Scrap Tires to Oil and Carbon Black by
      Vacuum Pyrolysis." *Resources, Conservation and Recycling.* 4:203-
      213; and U.S. Patent No. 4,740,270 (1988).

CHAPTER 3: POLLUTION PREVENTION SOLUTIONS

1.1. *Basic Concepts Solution-1.*

This unit conversion may be carried out as follows:

(200 ng/L) (1 g/10^9 ng) (1 lb/453.6 g) (3.785 L/1 gal)
= 1.67 x 10^{-9} lb/gal (10,000 gal/10,000 gal)
= 1.67 x 10^{-5} lb/10,000 gal

Alternatively, knowing 1 mg/L = 1 ppm = 8.34 x 10^{-6} lb/gal, the solution is:

200 ng/L (1 mg/10^6 ng) ((8.34 x 10^{-6} lb/gal)/(mg/L))(10,000 gal/10,000 gal) = 1.67 x 10^{-5} lb/10,000 gal.

1.2. *Basic Concepts Solution-2.*

The following are answers to the True/False questions:

1. True. S.T.P. does refer to conditions of 0°C and 1 atm. However, these are not the only S.T.P. conditions that are used in engineering practice.
2. False. Relative humidity is temperature dependent.
3. False. $C_2H_4 + 3 O_2 \rightarrow 2 CO_2(gas) + 2 H_2O(gas)$.
4. True. The Canadians use the Imperial gal (4.0 L) which is slightly larger than the U.S. gal (3.785 L)
5. True. The bomb calorimeter does give the "gross heating value" of a combustible material, not the "net heating value".
6. False. Excess air is defined as "air in excess of that required for stoichiometric combustion".
7. True. For ideal gas mixtures, the volume fraction = partial pressure in atm when the total pressure is 1 atm.
8. False. One gmole of SO_2 weights 64 g.
9. True. 100 ppmv = 0.01 vol%.
10. False. The density of water at 32°F is 62.4 lb/ft^3.

1.3. *Basic Concepts Solution-3.*

The molecular structure for each of these compounds is shown below:

1.3. Basic Concepts Solution-3 (continued).

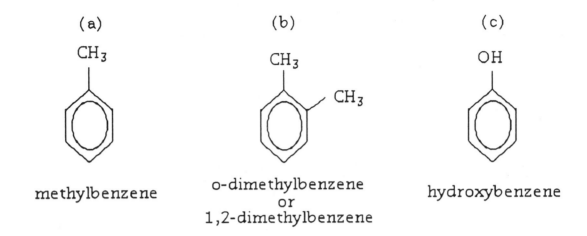

(a) methylbenzene

(b) o-dimethylbenzene or 1,2-dimethylbenzene

(c) hydroxybenzene

1.4. Basic Concepts Solution-4.

The concentration conversions for HCl, ethanol and ammonia are given in the table below:

	Concentration		
	% weight	ppm	molarity
HCl	36	562,500	15.40
Ethanol	3×10^{-4}	3	6.5×10^{-5}
Ammonia	1.67	17,000	1

1.5. Basic Concepts Solution-5.

Henry's law states

$$p_A = H \, x_A$$

where p_A = partial pressure of compound A (atm); x_A = mole fraction of compound A in solution (mol A/mol solution); H = Henry's constant (atm).

From the graph, the corresponding Henry's constants are:

Gas	H
H_2 at 20°C	68,300
H_2 at 30°C	72,900
C_2H_2 at 20°C	1,210
C_2H_2 at 30°C	1,460

1.5. *Basic Concepts Solution-5 (continued).*

The molecular weights for these compounds are also needed:

Gas	Molecular weight (g/gmol)
H_2	2
C_2H_2	26
H_2O	18

a. $x_{hydrogen}$ = 0.26 atm/68,300 atm
= 3.81 x 10^{-6} gmol H_2/gmol solution at T = 20°C.

The mole fraction of water is calculated as:

x_{water} = 1 - $x_{hydrogen}$ = (1 - 3.81 x 10^{-6}) gmol H_2O/gmol solution

dissolved H_2 = (3.81 x 10^{-6} gmol H_2/gmol solution) (2 g H_2/gmol H_2)/[(1 - 3.81 x 10^{-6}) gmol H_2O/mol solution (18 g H_2O /gmol H_2O)] = 4.2 x 10^{-7} g H_2/g H_2O = 0.42 mg H_2/L H_2O

b. For 30°C, a similar procedure leads to:

$x_{hydrogen}$ = 0.26 atm/72,900 atm
= 3.57 x 10^{-6} gmol H_2/gmol solution;

dissolved H_2 = 0.40 mg H_2/L H_2O.

c. $x_{acetylene}$ = 0.26 atm/1,210 atm
= 2.15 x 10^{-4} gmol C_2H_2/gmol solution at T=20°C.

x_{water} = 1 - $x_{acetylene}$ = (1 - 2.15 x 10^{-4}) gmol H_2O/gmol solution

dissolved C_2H_2 = (2.15 x 10^{-4} gmol C_2H_2/gmol solution) (26 g C_2H_2/gmol C_2H_2)/[(1 - 2.15 x 10^{-4}) gmol H_2O/gmol solution (18 g H_2O/gmol H_2O)] = 3.1 x 10^{-4} g C_2H_2/g H_2O = 310 mg C_2H_2/L H_2O.

d. For 30°C, a similar procedure leads to:

$x_{acetylene}$ = 0.26 atm/1,460 atm
= 1.78 x 10^{-4} gmol C_2H_2/gmol solution;

dissolved C_2H_2 = 257 mg C_2H_2/L H_2O.

1.6. *Basic Principles Solution-6.*

a. The block diagram for the process with its principal component parts is shown below.

1.6. Basic Concepts Solution-6 (continued).

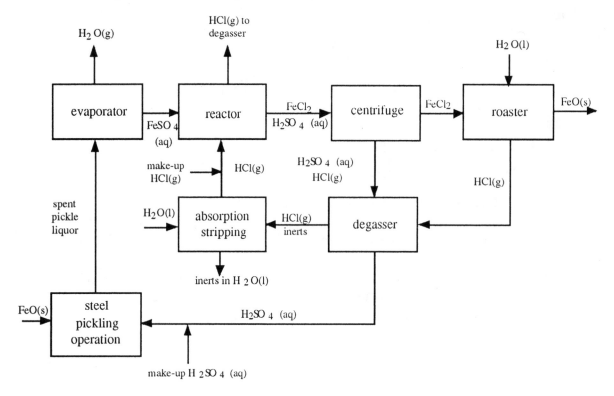

b. The overall reaction is written as the sum of the three primary reactions:

$$FeO + H_2SO_4 \rightarrow FeSO_4 + H_2O$$
$$FeSO_4 + 2\,HCl \rightarrow FeCl_2 + H_2SO_4$$
$$FeCl_2 + H_2O \rightarrow 2\,HCl + FeO$$

NET: $0 \rightarrow 0$

This scheme converts FeO(scale) to FeO(s) with minimum waste. Some waste will be created due to venting of streams to avoid impurity buildup, and the non-ideality of reactions taking place within the system.

The process is described in: D.F. Rudd, G.J. Powers, J.J. Siirola (1973), *Process Synthesis*, Prentice-Hall, Englewood Cliffs, New Jersey. (Problem 2.1).

1.7. Basic Concepts Solution-7.

The stream and industrial cooling requirement conditions can be graphically described in the figure below:

1.7. *Basic Concepts Solution-7 (continued).*

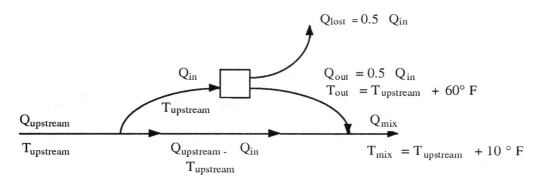

Writing a heat balance around the effluent yields the following equation, where Q = volumetric flow rate, C_p = heat capacity, and T = temperature:

$$C_p (Q_{upstream} - Q_{in}) (T_{upstream}) + 0.5 C_p Q_{in} (T_{upstream} + 60°F)$$
$$= C_p Q_{mix} (T_{mix})$$

C_p is assumed constant for all waste streams and can be ignored in further calculations.

Sustituting $Q = Q_{upstream}$ and $T = T_{upstream}$, and knowing that Q_{mix} = $(Q_{upstream} - Q_{in}) + 0.5 Q_{in}$ and $T_{mix} = T_{upstream} + 10°F$ yields:

$$(Q - Q_{in}) (T) + 0.5 Q_{in} (T + 60°F) = [(Q - Q_{in}) + 0.5 Q_{in}] (T + 10°F)$$

Multiplication yields:

$$Q T - Q_{in} T + 0.5 Q_{in} T + 30 Q_{in} = Q T - 0.5 Q_{in} T + 10 Q - 5 Q_{in}$$

Simplification and solving for Q_{in} yields:

$$Q T - 0.5 Q_{in} T + 30 Q_{in} = Q T - 0.5 Q_{in} T + 10 Q - 5 Q_{in}$$

$$Q_{in} = 0.286 Q$$

Therefore, 28.6% of the original flow, Q, is available for cooling to maintain the temperature constraints established in the problem.

1.8. *Basic Concepts Solution-8.*

Energy supplied by the coal (Q) = 100 MW produced (1 MW supplied)/(0.3 MW produced)
= [100 x 10^6/(0.3 J/s)] (0.2388 cal/J)] = 7.96 x 10^7 cal/s.

1.8. *Basic Concepts Solution-8 (continued).*

The coal requirement (F) = 7.96 x 10^7 cal/s (1 g coal/7,300 cal) (1 kg/1000 g)
= 10.9 kg/s coal (2.205 lb/kg) = 24.0 lb/s coal.

The amount of exhaust (E) = 45 lbmol/(100 lb coal) (24.0 lb/s coal)
= 10.8 lbmol/s.

To calculate the volume of exhaust in scfm, the ideal gas law is utilized:

PV = nRT

At standard conditions, P = 1 atm, T = 60°F = 520°R, R = 0.73 ft^3–atm/lbmol–R, and n = E = 10.8 lbmol/s.

V = 10.8 lbmol/s (0.73ft^3–atm/lbmol–R) (520°R)/1 atm
= 4099.7 ft^3/s = 4099.7 ft^3/s (60 s/min)
= 245,981 ft^3/min untreated gas.

Alternative solution:

1 lbmol of gases at S.T.P. (T = 60°F and P = 1 atm) occupy 379 ft^3.

V = 10.8 lbmol/s (379 ft^3/lbmol) (60 s/min)
= 245,592 ft^3/min untreated gas.

This solution is 0.16% smaller than the previous one.

1.9. *Basic Concepts Solution-9.*

Use a basis of 100 kg coal for all of the calculations. On this basis, the following weight and molar composition for the coal can be determined:

	kg	MW	kgmol
Hydrogen	5	2	2.5
Carbon	81	12	6.75
Sulfur	0.64	32	0.02
Nitrogen	1.4	28	0.05
Oxygen	5.76	32	0.18
Ash	6.2	- -	- -

The chemical reactions taking place within the incinerator can be described by the following stoichiometric equations:

1.9. *Basic Concepts Solution-9 (continued)*.

$$H_2 + 1/2\ O_2 \rightarrow H_2O \qquad\qquad (1)$$
$$S + O_2 \rightarrow SO_2 \qquad\qquad (2)$$
$$C + O_2 \rightarrow CO_2 \qquad\qquad (3)$$
$$1/2\ N_2 + O_2 \rightarrow NO_2 \qquad\qquad (4)$$
$$S + 1.5\ O_2 \rightarrow SO_3 \qquad\qquad (5)$$
$$C + 1/2\ O_2 \rightarrow CO \qquad\qquad (6)$$

Theoretical oxygen requirement based on the first three chemical reactions are: $1/2\ (2.5) + 0.02 + 6.75 - 0.18$
$= 7.84$ kgmol O_2/100 kg coal.

The amount of oxygen supplied with the air (15% excess)
$= 1.15\ (7.84$ kgmol $= 9.02$ kgmol O_2.

The amount of nitrogen with the air
$= 9.02$ kgmol O_2 (0.79 kgmol N_2/0.21 kgmol O_2)
$= 33.93$ kgmol N_2.

The product composition is as follows:

Product	kgmol	mol fraction
H_2O	2.5	0.056
CO_2	$0.995\ (6.75) = 6.716$	0.151
CO	$0.005\ (6.75) = 0.034$	7.6×10^{-4}
SO_2	$0.95\ (0.02) = 0.019$	4.3×10^{-4}
SO_3	$0.05\ (0.02) = 0.001$	2.3×10^{-5}
NO_2	$20\ \{100/[(1000)\ (46)]\} = 0.043$	9.7×10^{-4}
N_2	$33.93 + 0.05 - 0.043 = 33.937$	0.764
O_2	$0.18 + 9.02 - 2.5/2 - 6.716 - 0.034/2 - 0.019 -$ $0.001\ (1.5) - 0.043 = 1.154$	0.026
		$\Sigma = 0.999 \approx 1.0$

SO_x and NO_x along with particulates represent major sources of air pollution from thermal power plants. In this example SOx and NOx amount to approximately 400 to 1,000 ppm concentration in the flue gas and warrant further treatment.

1.10. *Basic Concepts Solution-10*.

a. The higher heating value per 1 kg of the coal fuel is estimated using DuLong's formula as follows:

1.10. *Basic Concepts Solution-10 (continued).*

HHV = 33801 (0.758) + 144,158 [0.051- 0.125 (0.082)] + 9413 (0.016)
= 31,646 kJ/kg coal = 31.65 MJ/kg coal = 0.03165 GJ/kg coal
= 13,617 Btu/lb coal.

b. Using a basis of 1 kg of coal, the heat released is:

Heat released = 0.03165 GJ.

1 kg of coal has 0.016 kg sulfur (S), which combusts according to the following stoichiometry:

$$S + O_2 = SO_2$$

This equation indicates that 1 gmol sulfur generates 1 gmol of SO_2, or 32.06 g S generate 64.07 g SO_2.

The SO_2 produced is:

0.016 kg S (64.07 kg SO_2/32.06 kg S) = 0.032 kg SO_2/kg coal.

SO_2 emissions = 0.032 (kg SO_2)/ 0.03165 (GJ) = 1.01 kg SO_2/ GJ
= 2.35 lb SO_2/MMBtu > 1.2 lb SO_2/MMBtu

c. The combustion of this coal is out of compliance. One can use pollution prevention measures to move back into compliance, i.e., use a lower sulfur and lower ash coal, oil, or natural gas. Another possibility is through air pollution control using, for example, a scrubber.

1.11. *Basic Concepts Solution-11.*

a. The balanced equations for the production of each sellable product are as follows:

$$2\ HNO_3 + Ca(OH)_2 \rightarrow Ca(NO_3)_2 + 2\ H_2O$$
$$2\ HF + Ca(OH)_2 \rightarrow CaF_2 + 2\ H_2O$$
$$2\ CH_3COOH + Ca(OH)_2 \rightarrow Ca(CH_3COO)_2 + 2\ H_2O$$

b. Using a basis of 1 lb of mixed acid, the lime requirement for each acid solution component is:

$$HNO_3 : \left(\frac{0.6\ lb\ HNO_3}{63\ lb/lbmol\ HNO_3} \right) \left(\frac{1\ lbmol\ Ca(OH)_2}{2\ lb\ mol\ HNO_3} \right) = 0.0046\ lbmol\ Ca(OH)_2$$

1.11. *Basic Concepts Solution-11 (continued).*

HF: $\left(\dfrac{0.2 \text{ lb HF}}{20 \text{ lb/lbmol HF}}\right)\left(\dfrac{1 \text{ lbmol Ca(OH)}_2}{2 \text{ lbmol HF}}\right) = 0.0050$ lbmol Ca(OH)$_2$

CH$_3$COOH: $\left(\dfrac{0.21 \text{ CH}_3\text{COOH}}{60 \text{ lb/lbmol CH}_3\text{COOH}}\right)\left(\dfrac{1 \text{ lbmol Ca(OH)}_2}{2 \text{ lbmol CH}_3\text{COOH}}\right) = 0.0017$ lbmol Ca(OH)$_2$

The total hydrated lime requirement is:

Ca(OH)$_2$:   0.0048 + 0.0050 + 0.0017 = 0.0115 lbmol Ca(OH)$_2$

On a mass basis, the total demand is:

$$\left(0.0115 \ \dfrac{\text{lbmol Ca(OH)}_2}{\text{lb acid}}\right)\left(74 \ \dfrac{\text{lb Ca(OH)}_2}{\text{lbmol Ca(OH)}_2}\right) = 0.85 \ \dfrac{\text{lb Ca(OH)}_2}{\text{lb acid}}$$

Problem motivation from C.G. Schwarzer (1973), *Environ. Sci. Technol.*, 13:166-171.

1.12. *Basic Concepts Solution-12.*

a. Oxidation-reduction reactions are most easily solved by separating the two operations into half-reactions and balancing those first.   The sodium ions do not enter into the reaction and are simply "spectator" ions.

Step one:

$\quad$ CrO$_4{}^{2-}$ $\rightarrow$ Cr^{3+} for the reduction half-reaction and

$\quad$ S$_2$O$_5{}^{2-}$ $\rightarrow$ SO$_4{}^{2-}$ for the oxidation half-reaction

Water and hydronium ions are conveniently used to add or subtract oxygen atoms in the balancing of the half-reactions.   Two H$_3$O$^+$ are required by this method to remove one oxygen atom, forming water. For oxidation in this problem, each oxygen atom added requires three hydronium ions.   (There are other methods for balancing of oxidation/reduction reactions which can be found in any general chemistry text.)

1.12. *Basic Concepts Solution-12 (continued).*

Step 2:

$$CrO_4^{2-} + 8\ H_3O^+ \rightarrow Cr^{3+} + 12\ H_2O$$

$$S_2O_5^{2-} + 9\ H_2O \rightarrow 2\ SO_4^{2-} + 6\ H_3O^+$$

Having achieved material balances in the half-reactions, a charge balance is the next consideration. Electrons, e⁻, are added or subtracted from each half-reaction to balance the charges on either side of the equations.

Step 3:

$$CrO_4^{2-} + 8\ H_3O^+ \rightarrow Cr^{3+} + 12\ H_2O - 3e^-$$

$$S_2O_5^{2-} + 9\ H_2O \rightarrow 2\ SO_4^{2-} + 6\ H_3O^+ + 4e^-$$

These equations simply state that the chromate ion is reduced, generating three electrons. At the same time, the metabisulfate ion is oxidized which requires four electrons. Obviously, the lowest common denominator is 12 and to totally balance the equation, on an element and charge basis, four chromate ions will exactly react with three metabisulfate ions.

$$4\ CrO_4^{2-} + 32\ H_3O^+ \rightarrow 4\ Cr^{3+} + 48\ H_2O - 12e^-$$

$$3\ S_2O_5^{2-} + 27\ H_2O \rightarrow 6\ SO_4^{2-} + 18\ H_3O^+ + 12e^-$$

Now the two half-reactions can be added together to yield:

$$4\ CrO_4^{2-} + 32\ H_3O^+ + 3\ S_2O_5^{2-} + 27\ H_2O \rightarrow 4\ Cr^{3+} + 48\ H_2O + 6\ SO_4^{2-} + 18\ H_3O^+$$

with the electrons cancelling out. Simplifying yields:

$$4\ CrO_4^{2-} + 3\ S_2O_5^{2-} + 14\ H_3O^+ \rightarrow 4\ Cr^{3+} + 6\ SO_4^{2-} + 21\ H_2O$$

This balanced equation shows that four moles of chromate ion will react with three moles of metabisulfite ion.

One lbmole of Na_2CrO_4 weighs 162 lb; therefore 5.0 lb Na_2CrO_4 is:

5.0 lb/(162 lb/lbmol) = 0.031 lbmol Na_2CrO_4

This amount will react with:

0.031 lbmol Na_2CrO_4 (3 lbmol $Na_2S_2O_5$/4 lbmol Na_2CrO_4) = 0.021 lbmol $Na_2S_2O_5$.

1.12. *Basic Concepts Solution-12 (continued).*

One lbmol of $Na_2S_2O_5$ has a mass of 190 lb; therefore 0.021 lbmol of $Na_2S_2O_5$ is:

0.021 lbmol $Na_2S_2O_5$ (190 lb/lbmol) = 4.0 lb $Na_2S_2O_5$

b. This metathesis reaction is easily balanced by the recognition that the cations are exchanging anions. Three OH^- ions are required for each of the two Cr^{3+} ions. Thus, multiplying the coefficient of the $Mg(OH)_2$ by a factor of 3 balances the equation on the left side.

$$Cr_2(SO_4)_3 + 3\ Mg(OH)_2 \rightarrow Cr(OH)_3 + 3\ MgSO_4$$

From this point the problem should be converted to metric units using gmol. From the equation found in the answer to Part a above, 4 gmol of chromate generate 2 gmol of $Cr_2(SO_4)_3$ upon reduction.

$$4\ CrO_4^{2-} \rightarrow 2\ Cr_2(SO_4)_3$$
(unbalanced, omitting non-essentials)

The stated amount of the chromate salt was 5.0 lb. The number of gmol of the sodium chromate is then:

5.0 lb (454 g/lb) = 2270 g; 2270 g/162 g/gmol = 14.0 gmol

This amount would then generate one-half the amount of $Cr_2(SO_4)_3$, according to the equation above, or 7.0 gmol. From the balanced equation in Part b, 1 gmol of the chromium (III) sulfate will react with 3 gmol of the magnesium hydroxide. Seven gmol of the sulfate then would require 21 gmol of the magnesium salt according to the following calculations:

7.0 gmol $Cr_2(SO_4)_3$ (3 gmol $Mg(OH)_2$/gmol $Cr_2(SO_4)_3$)
= 21 gmol $Mg(OH)_2$

One gmol of $Mg(OH)_2$ has a mass of 58.3 g; thus 21.0 gmol have a mass of 1224 g or 1.2 kg.

c. Molarity is a concentration term defined as the number of moles of a substance per L of solution. In the problem above, 5 lb of sodium chromate equals 14.0 gmol. This amount in 9.1 L of water would have a molarity of:

14.0 gmol/9.1 L = 1.54 M

1.13. *Basic Concepts Solution - 13.*

a. First, write a balanced stoichiometric reaction describing the complete combustion, and then add 25% excess air. Remember to carry the N_2 and O_2 to the product side of the equation. (The factor 3.78 in the term representing air results from the relationship: 0.791 lbmol N_2/0.209 lbmol O_2 = 3.78 lbmol N_2/lbmol O_2 in atmospheric air).

Stoichiometric Equation:

$$C_6H_5Cl + 7 (O_2 + 3.78\ N_2) \rightarrow 6\ CO_2 + 2\ H_2O + 26.46\ N_2 + HCl$$

25% Excess Air Equation:

$$C_6H_5Cl + 8.75 (O_2 + 3.78\ N_2) \rightarrow 6\ CO_2 + 2\ H_2O + 33.08\ N_2 + 1.75\ O_2 + HCl$$

b. First, the air-to-feed ratio is calculated on a molar basis. The molecular weights of each species are then computed and are used to convert results to a mass basis.

(8.75 lbmol O_2/lbmol C_6H_5Cl) (1 lbmol air/0.209 lbmol O_2) = 41.87 lbmol air/lbmol C_6H_5Cl

(41.87 lbmol air/lbmol C_6H_5Cl) (28.84 lb/lbmol air)/(112.45 lb/lbmol C_6H_5Cl) = 10.74 lb air/lb C_6H_5Cl

c. Two methods are available to solve this problem.

One method is similar to that used in Part b. The moles of each exhaust gas that are emitted per mole of fuel burned are computed using the balanced reaction from Part a. Then the total lb of exhaust leaving the stack is computed by multiplying the number of moles by the molecular weight for each species. Finally, the total lb of exhaust gas are divided by the lb of feed producing this exhaust and multiplied by the feed rate to obtain this exhaust feed rate.

The second employs a mass balance around the unit. The total mass input rates of the inlet streams must equal the total mass output rate through the stack (since complete combustion with no ash production is assumed). Then the ratio obtained in Part b can be used with the given fuel feed rate to generate the mass flow rate of gases leaving the stack.

1.13. *Basic Concepts Solution-13 (continued).*

Method 1: For every lbmol (112.45 lb) of fuel, the following mass of flue gases are generated

6 lbmol CO_2 (44 lb/lbmol)	= 264 lb CO_2
2 lbmol H_2O (18 lb/lbmol)	= 36 lb H_2O
33.08 lbmol N_2 (28 lb/lbmol)	= 926.2 lb N_2
1.75 lbmol O_2 (32 lb/lbmol)	= 56 lb O_2
1 lbmol HCl (36.45 lb/lbmol)	= 36.4 lb HCl
Total lbmol in Exhaust Gas	= 43.83 lbmol
Total Mass in Exhaust Gas	= 1318.6 lb total exhaust

The stack mass flow rate is then determined as follows:

(1318.6 lb exhaust gas)/(112.45 lb C_6H_5Cl) (100 lb C_6H_5Cl/hr) = 1173 lb exhaust/hr

Method 2: From Part b, the mass flow or exhaust air per lb of C_6H_5Cl is used to determine the stack mass flow rate as follows:

Mass flow rate of fuel + Mass flow rate of combustion gases = Mass flow rate of stack exhaust gases

(100 lb C_6H_5Cl/hr) (1 + 10.74 lb air/lb C_6H_5Cl) = 1174 lb exhaust/hr

d. Use the ideal gas law, remembering to convert to absolute temperature and pressure.

Molar flow rate of exhaust gases is determined as follows:

(43.83 lbmol exhaust gas/lb C_6H_5Cl) (1 lbmol C_6H_5Cl/112.45 lb) (100 lb C_6H_5Cl/hr) = 39.0 lbmol exhaust/hr

The volume of this exhaust gas is then calculated as follows:

V = nRT/P = (39.0 lbmol exhaust/hr) (0.7302 ft^3-atm/lbmol-°R) (860 °R)/(1 atm) = 24,490 ft^2/hr (1 hr/60 min) = 408.2 = 408 acfm

e. Use Charles' Law to determine the volume at standard conditions.

$V_2/V_1 = T_2/T_1$ ∴ V_2 = (408 acfm) (537 °R/860 °R) = 254.8 = 255 scfm

1.13. *Basic Concepts Solution-13 (continued).*

f. To convert to dry standard cubic feet, it is necessary to remove the water vapor from the volume of air as follows:

The dry lbmol of exhaust gases is the total lbmol of exhaust gas less the lbmol of water in the exhaust gas = 43.83 lbmol - 2 lbmol = 41.83 lbmol dry exhaust gas.

The ideal gas law can then be used to covert the volume to a dry basis as follows:

$$P_1V_1/P_1V_2 = n_1RT_1/n_2RT_2$$
$$\therefore \ V_1/V_2 = n_1/n_2$$

The dry volumetric flow rate is then calculated from the ratio of the wet to dry volumetric flow rates as:

Dry volumetric flow rate = (255 scfm) (41.83 lbmol dry exhaust)/(43.83 lbmol total exhaust)
= 243.4 = 243 dscfm

g. and h.   Simple unit conversions are required to answer this portion of the problem:

Inlet fuel: (100 lb/hr) (1 kg/2.2 lb) = 45.45 kg C_6H_5Cl/hr

Inlet air: (1074 lb/hr) ((1 kg/2.2 lb) = 488.2 kg air/hr

Exhaust mass flow rate: 45.45 kg C_6H_5Cl/hr + 488.2 kg air/hr
= 533.6 kg total exhaust/hr

Exhaust volumetric flow rates:
408 acfm (1 m^3/35.3 ft^3) = 11.66 = 11.7 m^3/min, actual

255 scfm (1 m^3/35.3 ft^3) = 7.2 m^3/min, standard conditions

243 dscfm (1 m^3/35.3 ft^3) = 6.9 m^3/min, dry, standard conditions

1.114. *Basic Concepts Solution - 14.*

a. To calculate the mass of oxygen required to completely oxidize a 1,000 gal spill of hexane in 1,000 yd^3 of soil requires first the determination of the mass of hexane spilled. Once this number is

1.14. *Basic Concepts Solution-14 (continued).*

calculated, the oxygen requirement of the hexane can be determined using the stoichiometric oxygen equivalent given in the problem statement. This number represents the maximum oxygen equivalent, and should be reduced by the fraction of the substrate that is incorporated into cellular material.

Mass of hexane to degrade = 1,000 gal C_6H_{14}/(7.48 gal/ft^3) [(0.659) (62.4 lb/ft^3)]
= 5498 lb C_6H_{14} (1 kg/2.2 lb) = 2499 kg C_6H_{14}

Maximum O_2 demand = 5498 lb C_6H_{14} (3.5 g O_2/g C_6H_{14})
= 19,243 lb O_2

Actual O_2 demand = 19,243 lb O_2 (1 - 0.2 cell incorporation)
= 15,394 lb O_2 = 15,394 lb O_2 (1 kg/2.2 lb) = 6997 kg O_2

b. The volume of this oxygen required for complete hexane biodegradation at STP (20°C, 1 atm) is calculated using the ideal gas law, PV = nRT, rearranged in the form V = nRT/P. The number of moles of oxygen is calculated as follows:

Moles of O_2 = 6997 kg O_2 (1000 g/kg)/(32 g/gmol O_2)
= 218,700 gmol O_2

Volume of O_2 = 218,656 gmol O_2 (R) (293 K)/(1 atm);
218,700 gmol O_2 (0.08205 atm-L/K-gmol) (293 K)/(1 atm)
= 5,257,000 L O_2

Volume of air = Volume of O_2/0.21 = 5,257,000 L/0.21 = 25,032,000 L
= 25,032,000 L/(28.31 L/ft^3) = 884,200 ft^3 of air

c. For the determination of the mass of available nutrient pool, the mass of soil requiring treatment is found from thespecific gravity of the soil and the contaminated volume:

Mass of soil = 1,000 yd^3 (27 ft^3/yd^3) (1.59) (62.4 lb/ft^3)
= 2,679,000 lb of contaminated soil

The soil available nutrient pool of 0.03 wt% P and 0.2 wt% N yields the following mass of P and N:

1.14. *Basic Concepts Solution-14 (continued).*

Mass of P = (0.0003 lb P/lb soil) (2,679,000 lb of contaminated soil) = 804 lb P

Mass of N = (0.002 lb N/lb soil) (2,679,000 lb of contaminated soil) = 5358 lb N

The mass of C in the C_6H_{14} is determined based on the concentration of C in hexane, i.e., 72 lb C/(86 lb C_6H_{14}), and the mass of hexane determined in Part a above,

Mass of C = 5498 lb C_6H_{14} [72 lb C/(86 lb C_6H_{14})] = 4603 lb C

Based on the C:N:P ratio of 100:20:1 on a weight basis, and assuming a nutrient application rate of 1.2 times stoichiometric, the following nutrient availability requirements must be satisfied during biodegradation:

Required N = lb C (20 lb N/100 lb C) = 4603 lb C (0.2 lb N/lb C) = 921 lb N (1.2) = 1105 lb N

Required P = lb C (1 lb P/100 lb C) = 4603 lb C (0.01 lb P/lb C) = 46.0 lb P (1.2) = 55.2 lb P

With these required nutrients available in the soil at levels of 5358 lb N and 804 lb P, these essential nutrients are well in excess. No nutrient addition would be required. However, the adequate mixing of moisture, nutrients, oxygen (the electron acceptor), and the waste constituents must take place or less than optimal reaction rates will occur.

1.15. *Basic Concepts Solution-15.*

a. The mass balance is simply F+A=E as indicated in this simple flow diagram:

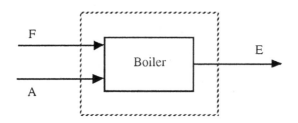

1.15. *Basic Concepts Solution-15 (continued).*

b. Since C is not combustible, $C_{in} = C_{out}$, so a component mass balance on C gives

$$yF + (0) A = zE$$

Using $E = F + A$, yields $yF = z(F + A)$, and $z = y (F/(F + A))$

Therefore, increasing the flow rate A decreases the mass FRACTION of C in the exhaust (a dilution effect). (This is the reason that <u>effluent</u> concentrations are given at a specified excess air concentration).

c. The new flow diagram can be drawn as indicated below:

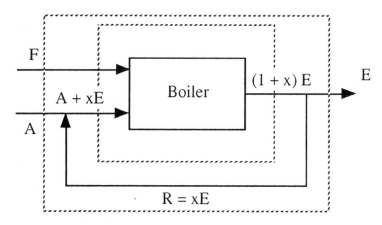

The overall mass balance is unchanged at:

$$F + A = E \text{ (unchanged)}$$

The mass balance around the reactor becomes:

$$F + A + xE = (1 + x) E$$
$$F + A = E \text{ (also unchanged)}$$

Note that we must be given the recycle amount (x) or some other information regarding concentrations to determine the recycle rate, since it cannot be determined from these equations alone.

d. The final revised flow diagram with the air pollution control unit takes the following form.

Note that the addition of the air pollution control unit requires another process stream, and that the pollutant has merely been transferred between media in this case.

1.15. *Basic Concepts Solution-15 (continued).*

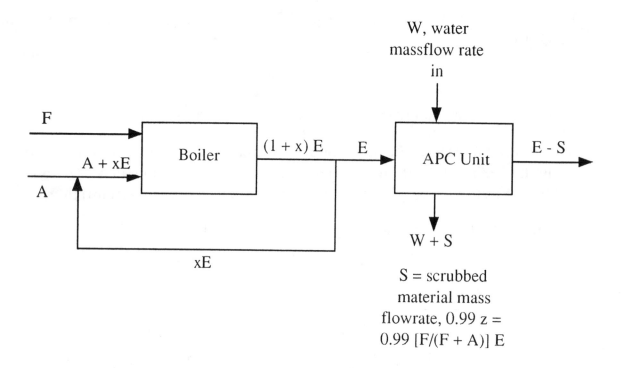

1.16. *Basic Concepts Solution-16.*

For Option A year one, the present value of the income is calculated as
follows:

$$PV = \$10,000 \ \frac{1}{(1 + 0.10)^1} = \$9091$$

For Option A year two the present value of the income is:

$$PV = \$15,000 \ \frac{1}{(1 + 0.10)^2} = \$12,397$$

The results for both options, all years are given in the table below.

From the table, it can be seen that Option A yields a higher present
value than Option B (\$45,145 versus \$45,079). Option A is the better
investment even though both options earn the same total
undiscounted income.

1.16. *Basic Concepts Solution-16 (continued).*

Table 2.   Return on investment for investment Options A and B expressed in Present Value terms.

Year	Annual Income Option A	Present Value Option A	Annual Income Option B	Present Value Option B
1	$10,000	$9,091	$10,000	$9,091
2	$15,000	$12,397	$10,000	$8,264
3	$10,000	$7,513	$15,000	$11,270
4	$10,000	$6,830	$15,000	$10,245
5	$15,000	$9,314	$10,000	$6,209
Total	$60,000	$45,145	$60,000	$45,079

This type of approach can be used to effectively evaluate different P^2 options.

2. Pollution Prevention Principles

2.1. *Pollution Prevention Principles -1.*

Pollution control is an end-of-pipe process. It is performed after the waste is generated. It often transfers the pollutant from one environmental medium to another.

Pollution prevention (P^2) focuses on the reduction of pollution at the source. This implies "up-front" limiting of waste.

P^2 was introduced in 1984 when the Congress included "waste minimization" in the Hazardous and Solid Waste Amendments of the 1984 RCRA law.

2.2. *Pollution Prevention Principles -2.*

1 - B
2 - D
3 - C
4 - A

2.3. *Pollution Prevention Principles -3.*

a. <u>Good housekeeping</u>: Improper labeling, storage and dumping of hazardous chemicals often increase the risk of spillage, ground water contamination and waste treatment plant upsets. Educating the employees on the proper handling and disposal of hazardous chemicals reduces such events tremendously.

b. <u>Material substitution</u>: This method involves replacing the polluting ingredients of a product with a less toxic one. Material substitution is most economical when a product is being developed for the first time. With an existing product, new chemicals often require an additional investment in new equipment. However, the additional investment may be less expensive than the cost of additional control systems necessary to meet new emission standards. For example, replacement of an organic based solvent as a carrier for tablet coating with a water based coating costs a pharmaceutical company $60,000. Alternatively, the air pollution control system to meet

2.3. *Pollution Prevention Principles -3 (continued).*

emission standards for the organic solvent had a capital cost of $180,000 and $30,000 annual operating costs. Clearly, this pharmaceutical company will replace the organic based coating with a water based one to realize the economic and pollution prevention benefit this material substitution provides.

c. Equipment design modification: Old equipment may contribute to the production of harmful by-products and excessive emission of pollutants. Replacing such process equipment might eliminate or minimize the generation of harmful by-products. Low interest federal loans may be available to assist in these changes.

d. Recycling: On-site waste recycle and reuse programs are very effective and reduce both raw material consumption and waste production. The use of off-site recyclers may be a practical option for smaller industries.

Some of the popular on-site recovery processes include: distillation, adsorption, filtration and electrolysis. Distillation is mainly used to recover up to 85 to 90 percent of the original organic solvent. The spent organic compound is distilled and the solvent vapor is condensed to recover the pure solvent. Adsorption processes generally use activated carbon beds to adsorb the solvent and can achieve up to 90 percent solvent recovery. The compound is recovered in a concentrated form during regeneration of the bed either by steam or inert gas regeneration. Filtration separates solids from the liquid stream via simple filtration or use of a fabric membrane. The separated solid and liquid streams could be reused with or without further treatment depending on their purity and their ultimate use. Electrolysis is commonly used for metal recovery from waste streams.

e. Waste exchange: Waste from one industry may actually be a valuable resource as a raw material in another industry. Avoided disposal costs for the waste generator and inexpensive raw material for the receiving industry are the inherent advantages of this waste reduction technique. Most of the chromium used in U.S. industries is imported. Recovery and reuse of chromium will not only minimize the amount of waste but also make the U.S. less dependent on imports.

2.3. *Pollution Prevention Principles -3 (continued).*

f. Detoxification: After considering all other options for waste reduction, detoxification of waste by neutralization or other techniques may be considered. Thermal, biological and chemical treatments are some of the detoxification processes commonly used. However, energy costs, along with the potential of generating new compounds and end product streams, must be evaluated thoroughly to justify the perceived environmental benefits of these proposed detoxification processes.

2.4. *Pollution Prevention Principles -4.*

A waste reduction plan must be specific to the industry. However, some general principles that must be incorporated for the success of the pollution prevention plan are discussed below.

Management commitment: Gaining the approval and support of top management is vital to the success of the waste reduction plan. It will be necessary to educate management about the pollution prevention program and its benefits through seminars and meetings. A short video tape and a brochure, such as the one used by 3M, may help. 3M successfully uses a 12-minute video on "Pollution Prevention Pays" to communicate the program to management and employees.

Communicating the Program to the Rest of the Company: Middle managers and employees with direct process line experience are in the best position to make suggestions as to where process improvements can be made. In addition, a monetary incentive and corporate recognition of employees for their practical pollution prevention ideas will also be very effective.

Performing a Waste Audit: The company must identify the processes, the products, and the waste streams in which hazardous chemicals are used. Mass balances of specific hazardous chemicals will help to identify waste reduction opportunities. Engineering interns could be very valuable in conducting such audits. Since managers and employees are often busy performing their assigned duties, an outside person may be able to focus on pollution prevention opportunities, cutting through some of the management and personnel barriers of the industry and may achieve significant progress.

2.4. *Pollution Prevention Principles -4 (continued).*

Cost/Benefit Analysis: Any change or modification in the process requires additional capital and operation and maintenance costs. A cost analysis must be included to help management make informed decisions. Factors including cost avoidance, enhanced productivity and decreased liability risks from the pollution prevention effort, should be factored into the study. Federal and State agencies provide matching grants to small industries to implement pollution prevention programs.

Implementation of Pollution Prevention Programs: Resistance to change by management and employees will still be an impediment for the implementation of the P^2 program. People known to resist new ideas and changes in the company must be included in the planning stages of the program. The CEO must be convinced of the merits of the program and must fully support the implementation of it.

Follow-Up: Reduced energy costs, reduced raw materials, and reduced waste disposal fees must be tracked and communicated to the company personnel. The tracking information will be very useful in the filing of the company's waste manifest and biennial reports on its waste reduction efforts required under new regulations. In addition, this information will show employees and management that pollution prevention programs not only make sense environmentally, but economically as well.

2.5. *Pollution Prevention Principles -5.*

a. Pollution prevention refers to the reduction or prevention of pollutant generation at the SOURCE. This concept was first defined as "waste minimization," but waste minimization can refer to methods that reduce the volume of waste after it is generated. In contrast, pollution prevention implies prevention of waste before it is generated.

Waste minimization was defined by the EPA in its 1986 report to Congress as:

"The reduction, to the extent feasible, of hazardous waste that is generated or subsequently treated, stored, or disposed of. It includes any source reduction or recycling activity undertaken by a generator that results in either: (1) the reduction of total volume or quantity of hazardous waste, or (2) the reduction of toxicity of hazardous waste,

2.5. *Pollution Prevention Principles -5 (continued).*

> or both, so long as such reduction is consistent with the goal of minimizing present and future threats to human health and the environment."

Pollution control refers to "downstream" reduction of pollution, i.e., treatment of process streams after waste has been generated. Frequently, pollution control measures simply transfer pollutants from one medium to another (for example, air pollutants to wastewater).

b. Examples of methods that might be used include
   1. Subsitution of a different, nontoxic or non–polluting solvent (e.g., citric acid–based) for a toxic solvent such as TCE. Note that when using a water soluble citric acid-based solvent, the grease and oil from the metal parts will float on top of the solvent and can be skimmed off so that the solvent can be re-used over and over again. This is an available option on many citric acid-based degreaser systems.
   2. Filtering and reuse of solvent for noncritical cleaning.
   3. Distillation of spent solvent to recover high–purity solvent that can be reused in the processing.
   4. Disposal of still bottoms from a solvent distillation process by shipping to an off–site landfill.

2.6. *Pollution Prevention Principles -6*

a. Specific answers will differ depending on the school examined and the research carried out in the various departments. Below are a set of typical answers for a small college setting.

The following departments and physical plant units are likely to be the major generators of waste on campus:

1. Chemistry Department - General chemistry labs (60-75%), mostly inorganic wastes (iodine, mercury, chromium, lead) and organic wastes (xylene, naptha, p-dichlorobenzene, and carbon tetrachloride). Organic Chemistry labs (25-40%), mostly organic solvents, some organic solids and chromium salts.

2.6. Pollution Prevention Principles -6 (continued).

2.   Biology Department - In the labs, formaldehyde, mercury salts, solvents and a small amount of osmium tetroxide are generated as wastes.

3.   Art Department - Acids, waste paint and solvents are generated from silk screen and printing processes.

4.   Psychology Department - Some waste solvents may be generated in animal research laboratories.

5.   Shop/Maintenance - Waste oils, paints, etc., may be generated from vehicle and/or equipment maintenance.

b.   Sample answers are suggested below. (This problem can best be used as a discussion question).

1.   Chemistry Department - To the extent possible, all labs should convert to microscale experiments. This is the single most productive way to avoid the generation of significant wastes AND to minimize student exposure. Experiments using toxic substances can often be conducted with alternative reagents (for example, the use of bleach instead of Chromium (VI) for oxidation reactions).

2.   Biology Department - Similar to the Chemistry Department.

3.   Art Department - Switch to water-base paints where possible.

4.   Psychology Department - Similar to the Chemistry Department. If animal research is not being carried out, then the amount of waste generated may be insignificant.

5.   Shop/Maintenance - Switch to water-base paints, avoid spills, recycle oil.

2.7. Pollution Prevention Principles -7.

a. The student should be able to write a paragraph which demonstrates his/her knowledge of the general characteristics of the various pollution scenarios listed in the problem statement.

2.7. *Pollution Prevention Principles -7 (continued).*

Suggested solutions are provided below:

i.   Soil at a railroad station contaminated with PCBs. Source reduction has already occurred since PCBs have been banned from production. Recycling is not applicable. One solutions is incineration (treatment), followed by landfilling (ultimate disposal). This is an actual real-life problem in the town of Croton-on-Hudson, NY.

ii. The Greenhouse Effect. The best solution, if there is one, is source reduction. This option has been recently discussed internationally at the Rio Conference in Brazil (June, 1992). There are a variety of sources of carbon dioxide, the leading cause of the Greenhouse effect. Combustion of fossil fuels, such as oil, coal, gasoline and wood, for energy production is one of the major sources of carbon dioxide emissions. Energy conservation, increasing energy efficiency, and use of alternative energy supplies will all contribute to a reduction in the use of fossil fuels for energy production. The destruction of the world's forests is also a contributing cause since plants utilize carbon dioxide in the formation of cellulose, generating oxygen. Recycling is not an option (except through photosynthesis as mentioned above), and treatment is extremely difficult since there are such vast quantities of gas requiring treatment. Ultimate disposal of gaseous carbon dioxide is not an option.

iii. Volatile organic compounds escaping from a stack. The best solution is to reduce the amount of VOCs used and lost by the process, i.e., source reduction. Recovery and recycling should be considered next; however, the effectiveness of this option is dependent on a number of variables, including the concentration of the VOCs being managed. The third best option would be treatment by incineration within the stack, or adsorption onto a bed of activated carbon where recycling and treatment options may be viable. Ultimate disposal of the spent carbon in a landfill will be the last choice option if allowed by current regulations.

iv. Ozone layer depletion. Again, the best hope lies in source reduction and in the development of alternative chemicals to replace chlorofluorocarbons (CFCs), the common refrigerants which are known to attack the ozone layer. This is the intent of the Montreal Protocol and later accords. Given the nature of CFCs and their interaction with the ozone layer, the problem will be with us for a

2.7. *Pollution Prevention Principles -7 (continued).*

long time. Treatment, recycling and ultimate disposal are not reasonable options for CFCs at the altitudes of concern. At ground level, recycling of refrigerants and ultimate disposal are activities which are actively being conducted. A major effort is directed toward finding more "ozone-layer friendly" substitutes for the CFCs. Unfortunately, the hydrochlorofluorocarbons (HCFCs) which are proposed as substitutes are still capable of damaging the ozone layer although to a much smaller extent than the CFCs.

v. <u>Radioactive wastes from a nuclear power plant</u>. There is only one solution for wastes already generated, and scientists are working to find the best means of ultimate disposal for this material. Leading the list of candidate treatment/disposal technologies is "glassification" whereby the radioactive wastes are imbedded in a stable glass form and placed in a geologically stable storage area. Since the half-life of many radioactive isotopes may be thousands of years, the best solution to future waste problems is source reduction. Radioactivity is not affected by any type of chemical change, so treatment of this waste is not a solution. Recycling of this waste to recover the plutonium and use this highly toxic substance in breeder reactors has been proposed. This process generates power and recovers plutonium. The United States has had a bad experience in the enrichment of radioactive wastes (West Valley, NY) and has decided to forego this option for the time being.

vi. <u>Chromium salts from printing/photographic industries</u>. The current solution for this waste is to reduce chromium (VI) to chromium (III) and then to precipitate it with base. The resulting solid is then ultimately disposed in a landfill. A better long-term solution is to reduce chromium use through source reduction. This can be accomplished in some instances by finding alternatives to chromium solutions. For instance, lab glassware was once routinely cleaned with a solution of chromium trioxide in sulfuric acid. The rinsings went down the sink. Today this is much less common, and unless absolutely crucial, an alternative cleaning agent is used. In many instances, chemists have found alternatives to chromium salts in oxidation reactions as well. Treatment alters chromium toxicity, but does not destroy the chromium. Recycling is generally too expensive to carry out relative to the cost of incorporating alternatives into current operations when feasible.

2.7. *Pollution Prevention Principles -7 (continued).*

vii. <u>Platinum catalyst in an industrial process.</u> Recovery and recycling is the best solution for this precious heavy metal. Economically, recovery and recycling is preferable to landfilling. Since it is used as a catalyst, source reduction can be accomplished by optimizing the design of specific reactors to minimize the amount of catalyst required. Alternative catalysts are heavy metals, so catalyst changes may not solve the problem. That which is not recycled should be landfilled.

viii. Sulfuric acid waste from the pharmaceutical industry. Usually the best option for these wastes is treatment with a strong base to form a salt and water at a pH of approximately 7. At this point the wastewater treatment plant can easily deal with the waste stream. In some favorable instances, the sulfuric acid can be cleaned, concentrated to its original molarity, and be reused. Depending on the application, source reduction may alleviate the problem, but the above options will be needed for whatever waste remains because ultimate disposal of an untreated waste stream is not an option.

b. The applicability of the axiom is fairly obvious and should pose no problem for the student. It is always more expensive to undo something than to have prevented the occurrence in the first place. There are a wealth of examples to choose from:

The Exxon Valdez Incident	Three Mile Island
Chernobyl	PCB pollution in the Hudson River
Lead contamination from paints	Pesticides in surface water
Smog in major cities: LA, NYC	Nuclear Weapons Plants
Love Canal	Strip Mining
Auto emissions, lead	Acid Rain
Mercury and cadmium in the Hudson River	
Algal "blooms" in ponds, lakes, and oceans	

2.8. *Pollution Prevention   Principles Solution - 8.*

a. The coal dust is a storage and handling emission. An accurate estimate of the amount of dust generated may be very difficult to make, as are suggestions for reducing this source. However, proper enclosure of the transfer point and exhausting the dust to a control system may minimize the emission rate. This dust contributes to

2.8. *Pollution Prevention Principles -8 (continued).*

the ambient particulate loading and as such is considered a pollutant. [Emission factors exist for this source].

b. The evaporating solvent is classified as a fugitive emission. An estimate of the magnitude of this emission can be made by comparing volumes of fresh and spent solvent in the bath. One possibility is for the development of an automated system that removes the lid only for purposes of inserting or removing parts. It may also be possible to reduce evaporation by temperature control of the bath (providing the bath is effective at reduced temperatures) or of the air space above the solvent liquid surface, i.e., a vapor condenser. Another option might be to install an exhaust hood to remove fumes, which can then be treated (for example, adsorbed before the air is vented to the atmosphere). The evaporated solvent is considered an air pollutant.

c. The spent solvent is a process emission and may be classified as a hazardous waste, depending upon its chemical composition. The amount of solvent waste may be reduced by recovering the solvent and reusing it (via distillation or other appropriate methods). It may be possible to replace the solvent with one that is not environmentally hazardous.

d. The solvent emitted from the stack (as a gas) is a process emission, and is classified as a pollutant. The amount of solvent lost in this way may be reduced by condensation of the solvent and separation from the stack gas.

The aerosol is classified as a secondary pollutant. This term refers to pollutants that result from further reaction of primary emissions; frequently, secondary pollution occurs at different locations (downwind, for example) away from the primary emission. The production of secondary pollutants depends upon a number of ambient conditions, primarily ambient volatile organic levels, NOx levels, temperature, and solar irradiation. In general, smog levels can be reduced by a concerted program of emissions abatement from automobiles and industry, leading to reduced ambient concentrations.

2.8. *Pollution Prevention Principles -8 (continued).*

e. The "spill" occurs as a part of the processing of the piece and can be classified as a process emission. Depending upon the chemicals used in the bath, the water used to wash the spill down the drain may become a waste or hazardous waste stream. It may be possible to reduce the problem by changing the location of the rinse bath so that the path of the piece remains over collection tanks, and by suspending the piece for a longer amount of time over the bath to allow most of the bath solution to drip back into the tank.

f. The leaked solvent is a fugitive emission. An estimate of the amount of solvent lost as fugitive emissions can be made by performing a material balance for the solvent over the entire process. Checking the integrity of the pump may reveal a reparable leak, or it may be necessary to switch to a different type of pump that is less likely to leak. It may also be possible to reduce this type of loss in process piping (at gaskets, for example) by shrink–wrapping the joint or otherwise containing the vapor. If possible, another solvent might be considered that has a lower vapor pressure or that is less corrosive to the pump interior.

g. If the acidic stream has a pH < 2, it is classified as a hazardous waste and must undergo further treatment before disposal. Similarly, if a caustic effluent has a pH > 12, it is classified as a hazardous waste. It may be possible to replace the acidic cleaner with one that is not classified as hazardous. It may also be possible to recover spent solution and recycle it for cleaning purposes, reducing the amount of waste generated.

h. If either the acidic or caustic stream is classified as hazardous (see g), the neutralized stream may also be considered hazardous even if its pH is greater than 2 and less than 12. This is known as the "derived–from" rule, which assigns the hazardous classification to derivatives from hazardous wastes. It may be possible to reduce the amount of neutralized solution to be processed by concentrating it; the water thus recovered may be recycled for use in other parts of the process.

2.9. *Pollution Prevention Principles Solution - 9.*

The nitrogen allows the solvent vapor to exist at much higher concentrations without explosion. Consequently the vapor could be recovered more effectively by condensation. In addition, this may save energy.

Nitrogen-based systems are much safer from an explosion point of view than conventional air systems. A nitrogen system can be built as a self-contained closed system minimizing the discharge of any solvent vapor to the atmosphere. A typical block diagram of such a system is shown in below.

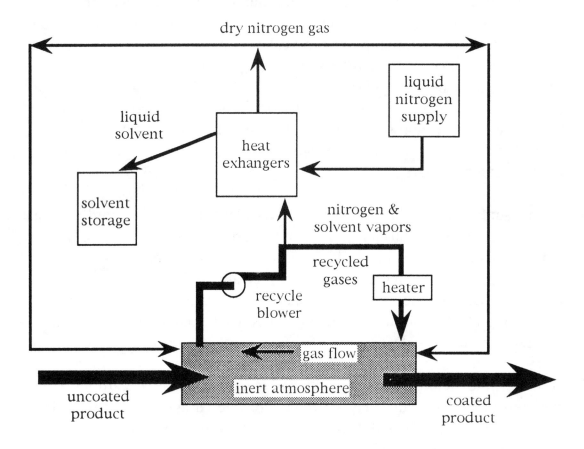

Figure 1. Schematic of a closed-loop nitrogen-based drying oven.

2.10. *Pollution Prevention Principles Solution - 10.*

There is, of course, no single answer to this question. For an excellent discussion of the subject refer to Holmes, G., B. R. Singh, and L. Theodore (1993), *Handbook of environmental management*, John Wiley and Sons, Inc., New York. Below are some examples of "correct" answers.

2.10. Pollution Prevention Principles -10 (continued).

A.   B.

1. Took a long 15 min. shower.   1. a. Take short showers.
                                          b. Use a restricted shower head.

2. Used a fresh towel to dry after   2. Could have used yesterday's towel
   showering.   one more time.

3. Used a full flush of water in   3. a. Could let it "mellow" if yellow.
   toilet (6X).   b. Could use a reduced volume to
                                            flush.

4. Left water on while brushing   4. Turn off water.
   teeth.

5. Hot water tank set to 150°F.   5. Unnecessarily high. Set to
                                         lower temperature, i.e., 120°F.

6. Drove to store for   6. a. Walk or bicycle to nearby
   groceries (2X)   store.
                                          b. Consolidate trips.
                                          c. Use mass transit.

7. Used plastic/paper bags for   7. a. Take back your old bags for
   groceries, etc.   reuse.
                                          b. Use a non-disposable bag.

8. Left lights (TV, radio, etc.) on   8. a. Turn off lights when not used.
   in an empty room.   b. Install motion sensors.

9. Used incandescent bulbs.   9. Switch to compact fluorescent
                                         bulbs.

10. Opened the refrigerator to   10. a. Organize refrigerator so you
    search for something to eat   know where/what is there.
    (4X).   b. Determine what you want
                                            before opening door.
                                          c. Buy an energy-efficient model.

11. Drove to school/work   11. a. Walk.
    alone.   b. Set up a carpool.
                                          c. Use mass transit.
                                          d. Use a bicycle.

2.10. *Pollution Prevention Principles -10 (continued).*

A.

B.

12. Car gets 15 mpg.

12. Buy a car which gets higher gas mileage, i.e., >30 mpg.

13. Freezer in refrigerator is empty.

13. It is more efficent to store the (negative) heat in extra ice cubes, frozen bottles of water, etc.

14. Threw banana peel in trash.

14. Compost all suitable wastes.

15. Mowed lawn with power mower

15. a. Use a push, reel-type mower
    b. Buy or rent a goat.
    c. Consider a lawn covering plant in suitable areas.

16. Discarded lawn clipping in trash.   the lawn clippings.

16. Compost or grasscycle (mulch)

17. Heated tea kettle on gas stove for only one cup of tea.

17. Use a microwave oven for one cup. Avoid reheating the tea/coffee.

18. Washed a few clothes.

18. a. Wash full loads of clothes.
    b. Wear clothes one more time.

19. Dried a few clothes.

19. a. Dry a full load of clothes.
    b. Sun-dry on a clothes line.

20. Run heat all day in winter.

20. a. Install a timed thermostat to heat only when people are at home.
    b. Use sweaters and warm clothes and turn down the thermostat.

21. Tossed spent AA batteries in trash.

21. a. Do not throw old batteries in trash. Many places now collect batteries for return to the manufacturer.
    b. Purchase rechargeable batteries and a battery charger.

2.11. *Pollution Prevention Principles Solution - 11.*

Data needed for a quantitative analysis of this problem must be in the form of relative emission rates per unit mass transported. Emission data for loading and unloading activities should also be provided.

a. Merits of truck tansportation

1. Trucks can usually travel a more direct route for a delivery (fewer miles traveled) than trains. This is due to the fact that the infrastructure of roads and highways is much more extensive than that of the railroads. In many cases, a truck will carry an entire load to a single destination, taking the most direct route. Because the railroad is limited by its infrastructure, trains will most likely have to travel farther to reach the same destination.

2. Truck engines produce fewer emissions than railroad engines. Beginning in 1991, the EPA set emission standards for all classes of heavy-duty over-the-road trucks. The emission standards require heavy-duty diesel engine manufacturers to decrease engine emissions of particulates (carbon soot), hydrocarbons, NOx, and CO. To meet these standards, engine manufacturers had to redesign many engine components and develop better combustion technologies. So far, the EPA has not set emission standards for railroad diesel engines. Railroad diesels are typically much larger than truck diesels and are built by different manufacturers. Because there are no emission standards for the railroad diesels, they continue to be built with less expensive, obsolete engine technology that produces higher emissions. To compare exhaust emissions on an equal basis, they need to be measured in grams per hour per horsepower produced, or tons of load delivered per kg of pollution per mile.

b. Merits of railroad transportation

1. Unlike trucks, trains can couple several hundred cars together and pull them with only a few large diesel engines. Because of this, trains accelerate very slowly but are able to reach and maintain crusing speeds (analogous to a large, heavy car with an undersized engine). Because a train has a much lower power-to-weight ratio than a truck, it will get better fuel economy. To compare a train and truck on an equal basis, the fuel economy (mpg) should be calculated

2.11. *Pollution Prevention Principles -11 (continued).*

for both and divided by the total weight carried, resulting in mpg per ton values.

2. Trains travel at steadier speeds than trucks. Fuel economy and emissions are optimized during steady-state operating conditions with an internal combustion engine. Trucks are exposed to much more stop-and-go driving than trains during which the engine is repeatedly accelerated and decelerated. Engines get poor fuel economy during acceleration because more fuel must be used to prevent them from "bucking" or "stalling." Likewise, emissions are much higher during acceleration due to the overfueling that takes place. With a diesel engine, one can actually see the emissions (particulates) bellowing from the exhaust during acceleration.

3. Another section could be added discussing the proper balance of trucking and railroad use. Some would say that we have gone too far from the optimum in the direction of trucks or buses and make too little use of trains. This discussion could include an analysis of the optimal mix of trucking and trains to deliver goods in a fashion that maximizes pollution prevention.

3. Regulations

3.1. *Regulations Solution - 1.*

The following are some of the major differences between a federal *law* and a federal *regulation*.

1. A law is passed by both houses of Congress and signed by the President. A regulation is issued by a government agency such as the U.S. Environmental Protection Agency (U.S. EPA) or the Occupational Safety and Health Administration (OSHA).

2. Congress can pass a law on any subject it chooses. It is only limited by the restrictions in the Constitution. A law can be challenged in court only if it violates the Constitution. It may not be challenged if it is merely unwise, unreasonable or even silly. If, for example, a law were passed that placed a tax on sneezes, it could not be challenged in court just because it was unenforceable.
   A regulation can be issued by an agency only if the agency is authorized to do so by the law passed by Congress. When Congress passes a law, it usually assigns an administrative agency to implement that law. A law regarding radio stations, for example, may be assigned to the Federal Communications Commission (FCC). Sometimes a new agency is created to implement a law. This was the case with the Consumer Product Safety Commission (CPSC). OSHA is authorized by the Occupational Safety and Health Act to issue regulations that protect workers from exposure to the hazardous chemicals they use in manufacturing processes. If those hazardous chemicals are emitted by the plant and affect the surrounding community but do not expose the workers in the plant, OSHA is not authorized to issue an order to stop the practice (Note: the U.S. EPA is authorized to regulate such practice).
   A regulation may be challenged in court on the basis that the issuing agency exceeded the mandate given it by Congress. If the law requires the agency to consider costs versus benefits of their regulation, the regulation could be challenged in court on the basis that the cost/benefit analysis was not correctly or adequately done. If OSHA issues a regulation limiting a worker's exposure to a hazardous chemical to 1 part per million (ppm), OSHA could be called upon to prove in court that such a low limit was needed to prevent a worker from being harmed. Failure to prove this would mean that OSHA exceeded its mandate under the law, as OSHA is charged to

3.1. *Regulations Solution - 1 (continued).*

develop standards only as stringent as those required to protect worker health and provide worker safety.

3. Laws are usually brief and general. Regulations are usually lengthy and detailed. The Hazardous Materials Transportation Act, for example, is only approximately 20 pages long. It speaks in general terms about the need to protect the public from the dangers associated with transporting hazardous chemicals and identifies the Department of Transportation (DOT) as the agency responsible for issuing regulations implementing the law. The regulations issued by the DOT are several thousand pages long and are very detailed down to the exact size, shape, design and color of the warning placards that must be used on trucks carrying any of the thousands of regulated chemicals.

4. Generally, laws are passed infrequently. Often years pass between amendments to an existing law. A completely new law on a given subject already addressed by an existing law is unusual. Laws are published as a "Public Law #__-___" and are eventually codified into the United States Code.
Regulations are issued and amended frequently. Proposed and final new regulations and amendments to existing regulations are published daily in the Federal Register. Final regulations have the force of law when published. Annually, the regulations are codified in the Code of Federal Regulations (CFR). The CFR is divided into 50 volumes called Titles. Each Title is devoted to a subject or agency. For example, labor regulations are in Title 29, while environmental regulations are in Title 40.

3.2. *Regulations Solution - 2.*

This is an open-ended question. More than one correct answer is possible.

a. The Pollution Prevention Act of 1990 has no provisions related to enforcement and contains no penalties for failure to comply.

b. The effectiveness of implementation of the Pollution Prevention Act of 1990 relies heavily on voluntary compliance as indicated below:

3.2. *Regulations Solution - 2 (continued).*

i. The Pollution Prevention Act of 1990 calls on companies to disclose and report a great deal about their operations. Widespread inspections to determine compliance would be very expensive. It would also severely strain the government's manpower.

ii. The law aims at creating a more cooperative relationship between the environmental agencies and industry. Strict enforcement provisions with penalties for incomplete compliance could do the opposite and actually create a disincentive to critical self-auditing, self-policing and voluntary disclosure.

iii. Penalties for willful non-compliance, however, could be effective in reducing any open flaunting of the law.

iv. Companies have an incentive to voluntarily comply with the law because having smaller quantities of chemicals to dispose of could actually save money while giving the company a public relations edge.

3.3. *Regulations Solution - 3.*

The description of the major thrust and the choice of goals and provisions given for each law is subjective and therefore your answers may not exactly match those given here.

a. The National Environmental Policy Act (NEPA) requires any federal agency proposing a project that might affect the environment to prepare an environmental impact statement. The statement describes the project's adverse effect on the environment as well as its benefits. The statement also includes a discussion of alternatives.

b. The Resource Conservation and Recovery Act (RCRA) aims primarily to protect the environment from the mishandling of the land disposal of municipal, industrial, commercial and hazardous wastes. A major part of the law relevent to pollution prevention is Subtitle C, regulations pertaining specifically to hazardous waste management. A waste is hazardous if it appears on a designated list or possesses certain defined characteristics (ignitability, corrosivity, reactivity, and/or extraction procedure toxicity) as determined by prescribed tests. Hazardous waste generators, transporters, and hazardous waste treatment/storage/disposal facilities are required to keep records, submit reports, meet standards set forth in

3.3. *Regulations Solution - 3 (continued).*

regulations, and to obtain permits. RCRA attempts to create a cradle-to-grave hazardous waste management system.

c. <u>The Comprehensive Environmental Response, Compensation and Liabilities Act (CERCLA or "Superfund")</u> attempts to provide the means by which abandoned hazardous waste sites may be remediated. A special tax is placed on certain chemicals to raise the needed funds. A list of hazardous waste sites needing clean-up (The National Priorities List) is generated through an evaluation of the relative risk to public health and the environment posed by each site. The U.S. Environmental Protection Agency is given the authority to seek reimbursement for the clean-up expenses from the owners/operators of the site and from the generators of the waste.

d. <u>The Superfund Amendments and Reauthorization Act of 1986 (SARA) Title III</u> - also known as The Emergency Planning and Community Right-to-Know Act, requires each state to establish local emergency planning districts and committees to develop, together with certain chemical facilities, contingency plans for handling and responding to the release of an "Extremely Hazardous Substance." The affected facilities are those that have on their premises any chemical found on a list of "Extremely Hazardous Substances" published by the U.S. Environmental Protection Agency in an amount known as the "Threshold Planning Quantity" which is given on the same list.
Such facilities must submit an inventory of these chemicals and also report releases of reportable quantities of these chemicals to the emergency planning committee. Facilities must also account for the total quantity of substances brought in versus the amount shipped out (mass balance study). The difference could indicate a release of the chemical. The U.S. Environmental Protection Agency is to maintain a database of release reports which is available to the public.

e. <u>The Pollution Prevention Act of 1990</u> seeks to make the reduction of the production of pollutants ("source reduction") the primary means for controlling pollution (as opposed to treatment, storage, and disposal). The U.S. Environmental Protection Agency is to establish a source reduction program which: collects and disseminates information; provides financial assistance to states; establishes

3.3. *Regulations Solution - 3 (continued).*

standard methods for measuring source reduction; provides a source reduction clearinghouse; creates an advisory panel of experts from government, industry, and public groups; establishes training and award programs; promotes a multi-media approach to source reduction; and establishes a special office to oversee the implementation of this law.

3.4. *Regulations Solution - 4.*

a. D002. Sodium hydroxide solution is corrosive.

b. D001 and D003. This waste is both ignitable and extremely reactive with the potential to explode if exposed to air.

c. D004. PCBs are classified *toxic* by law.

3.5. *Regulations Solution - 5.*

Small Quantity Generators (SQGs), generate between 100 and 1000 kg HW/calendar month. Very Small Quantity Generators (VSQGs) generate less than 100 kg HW/calendar month.

SQGs generate substantial amount of waste (about 940,000 metric ton/yr); yet they have fewer than five or 10 employees, and the managers may have limited training in hazardous waste management.

Examples of SQGs include photography, printing, dry cleaning, retail establishments, etc. Example of their major wastes include spent solvents, strong acids, ignitable wastes, etc.

3.6. *Regulations Solution - 6.*

The "mixture" rule classifies any mix of a hazardous waste and a nonhazardous waste as a hazardous waste. It states that any solid waste mixed with a listed hazardous waste becomes a hazardous substance and is to be handled as hazardous.

3.6. *Regulations Solution - 6 (continued).*

The "derived from" rule classifies any residue resulting from the treatment of a hazardous waste as a hazardous waste. Any waste "derived from" the treatment of a listed hazardous waste remains a hazardous waste. An example of the "derived from" rule is ash from a hazardous waste incinerator, which carries the same waste code as the original waste that was incinerated.

Pollution prevention concepts apply to these rules from the standpoint of preventing the production of wastes that may be classified as hazardous waste. The practice of waste segregation, which isolates incompatible wastes to prevent unwanted reactions, and optimizes waste treatment, would also aid in minimizing the potential for application of the "mixture" rule to large waste volumes. Prevention of hazardous waste generation via product substitution, process substitution, modification of reaction chemistry, etc., would be necessary to avoid having the material classified as hazardous according to the "derived-from" rule.

3.7. *Regulations Solution - 7.*

The regulations and standards as given in 40 CFR Part 60 Subpart GGG define fugitive emission terms, sources and remediation technology. Part 60 was amended by adding Subpart GGG (Reference 1) for petroleum refineries. The standards directly list pumps, compressors, pressure-relief devices, valves, sampling connections, and vent and control devices. Test methods, record keeping and reporting requirements are specified. Part 60, Appendix A describes the leak detection methods (Method 21).

One major source of fugitive emissions is categorized as equipment leaks resulting from incomplete sealing at the point where liquid interfaces with the environment. These leaks include fugitive emissions from devices cited in the regulations: pumps, valves, pressure relief valves, open-ended lines, compressors, and sampling connections. The primary remediation approach is to aggressively test for leaks and to stop them through routine maintenance and, in some cases, replacement with specially designed pumps, seals, etc.

3.7. *Regulations Solution - 7 (continued).*

The reader should list the devices, summarize the leak test frequency and describe the allowable leak rates and frequencies. Method 21 should be mentioned briefly to indicate leak detection quantities.

3.8. *Regulations Solution - 8.*

EPA's "Derived-From" and "Mixture" Rules under RCRA state that any mixture of a listed hazardous waste and a nonhazardous solid waste is itself a RCRA hazardous waste and that any waste derived from the treatment, storage, or disposal of a listed waste is deemed hazardous. Both the "derived-from" and "mixture" rules apply regardless of how small of the concentration of the listed waste ("the solution to pollution is not dilution") in the final mixture. Likewise, mitigation of the characteristics which prompted the waste to be listed in the first place does not magically purify the material. In the case of listed hazardous wastes, the only method of removing the taint is by the EPA's delisting procedure, a process that has only rarely been successfully used.

As the employee is leaving, you might suggest that he or she read Jeffrey Gaba's excellent interpretation of the rules as published in the *Environmental Law Reporter*, January 1991. pp. 10022-10044.

3.9. *Regulations Solution - 9.*

A discussion of the provisions of each Title of the Clean Air Act Amendments of 1990 can be found in Holmes, G., B. R. Singh, and L. Theodore (1993), *Handbook of Environmental Management and Technology*, John Wiley and Sons, New York.

The Clean Air Act Amendments of 1990 are also cited as Public Law 101-549, 42 United States Code Section 7401 et seq. The text of the law may be found in the United States Code which many libraries possess. The text of the law may also be found in *Environmental Statutes*, 1992, or later published by Government Institutes, Inc., Rockville, MD.

3.9. *Regulations Solution - 9 (continued).*

The following is a brief summary of some of the issues addressed by, and major provisions of, each of the first seven Titles of the Clean Air Act Amendments of 1990. As the Act is very lengthy and complex, your choice of issues and provisions to discuss may not be identical to those given below.

Title I. Title I addresses the attainment of prescribed levels of clean air for specific gaseous pollutants and particulates. These levels of clean air are called "National Ambient Air Quality Standards" (NAAQSs). Each area of the nation will be classified as either an "Attainment" area (having met the NAAQSs), or a "Nonattainment" area (not having met the NAAQSs). The nonattainment areas will be further classified according to how bad the quality is in that area. The worse the air quality, the more stringent the actions that must be taken by the states and the U.S. EPA to improve air quality. Some of the actions that can be taken include: requiring facilities to obtain permits, imposition of "Reasonably Available Control Measures" (RACMs), and in some severe cases imposition of the "Best Available Control Measures" (BACMs). Title I also lowers the amount of emitted NAAQS pollutants that result in a facility being classified as a "Major Source." This section of the law also requires U.S. EPA to assist industry by issuing "Control Technique Guidelines" (CTGs) for various pollutants.

Title II. Title II addresses air pollution from mobile sources, i.e., cars, buses and trucks. New standards for car tailpipe emissions will be phased in starting with the 1994 model year. Manufacturers will be required to reduce the evaporation of gasoline during refueling, oxygenated and alcohol containing fuels will be required in certain nonattainment areas in the winter months, urban buses and centrally fueled fleets will have to meet special requirements, and an experimental pilot program for reducing emissions from cars will be instituted with the 1996 model year in California.

Title III. Title III addresses air pollution by toxic chemicals not cited in the NAAQSs. The release reporting requirements of Section 313 of Title III of the Superfund Amendments and Reauthorization Act (SARA) of 1986 (also known as the Emergency Planning and Community Right to Know Act) revealed that many such chemicals are emitted in large quantities. Estimates are that as many as 3,000

3.9. *Regulations Solution - 9 (continued).*

deaths may be attributed to these air pollutants each year in the United States.

Title III lists 189 toxic air pollutants which the U.S. EPA must regulate. The U.S. EPA must organize the sources of these pollutants into categories. Each category will include "Major Sources" (those that emit 10 T/yr of any one pollutant or 25 T/yr of any combination of these pollutants) and "Area Sources" (smaller sources such as dry cleaners). The U.S. EPA must then issue "Maximum Achievable Control Technology" (MACT) standards for each source category.

Title III also requires that the U.S. EPA establish new standards for municipal, industrial, and medical waste incinerators.

Title IV. Title IV addresses the problem of acid rain caused by the emission of sulfur dioxide (SO_2) and nitrogen oxides (NO_x). The primary emitters of these pollutants are coal-fired electric power plants.

Power plants will be required to reduce their emissions of SO_2 and NO_x in phases beginning in 1995. A system of "Allowances" will allow existing plants to buy the right to emit a certain quantity of these pollutants from new power plants that emit much less pollutant than the law allows. It is hoped that the ability to sell allowances will encourage the construction of very clean new power plants, and result in a lower total emission rate nationwide. Power plants that neither reduce their emissions adequately, nor buy sufficient allowances will have substantial fees imposed on them based on the quantity of their emissions.

Title V. Title V addresses the need to assure compliance with the new air pollution regulations. While the pre-1990 Clean Air Act required permits only for new stationary sources, the 1990 amendments require operating permits for existing sources as well. As before, the states develop the permitting program. Under the new amendments, however, the U.S. EPA will be able to oversee the state programs to insure that they are adequate. If a state fails to develop a permitting program, then the U.S. EPA must take over from the states.

Title VI. Title VI addresses the need to protect the ozone layer from chlorofluorocarbons (CFCs), and to reduce the threat of global warming. CFCs are to be completely phased out of production and

3.9. *Regulations Solution - 9 (continued).*

use. Some CFCs will be banned as early as the Year 2000. The U.S. EPA must publish a list of safe and unsafe CFC substitutes. The unsafe substitutes must also be banned. Non-essential products that release CFCs are to be banned as well by 1992. Aerosols and non-insulating foams using CFCs or hydrochlorofluorocarbons (HCFCs) must be banned by 1994 (with certain exemptions related to flammability and safety).

Title VII. Title VII deals with enforcement. The dollar amount of penalties that the U.S. EPA can impose on violators has been substantially increased above previous levels. Criminal penalties for knowingly violating the law have been raised from misdemeanors to felonies. Pollutant sources must certify their compliance with the regulations. The U.S. EPA can issue subpoenas for compliance data. The law provides for the payment of a bounty of up to $10,000 for citizens who furnish information leading to a conviction or penalty, and allows citizen groups to sue for past violations of the Act. Penalties arising from citizen suits go into a special fund to be used by the U.S. EPA for compliance and enforcement activities.

3.10. *Regulations Solution - 10.*

The following are some of the criteria used by the Department of Justice to assess the potential for criminal prosecution. You may have thought of some that are not listed here.

1. Was the violation discovered by an internal audit or by a government investigation? Did the company commit adequate resources to its internal self-audit program? Does the self-audit program appear likely to uncover or prevent future violations? Does the self-audit program contain safeguards to ensure its integrity?
2. Assuming the violation was discovered by the company, was it disclosed
   a. voluntarily
   b. promptly upon discovery
   c. with sufficient data of good quality
   d. before the regulatory agency discovered it?
3. What was the extent and quality of the cooperation with any government investigation of the violation?

3.10. *Regulations Solution - 10 (continued).*

4. Was the violation an isolated incident or one of many violations? How serious was the violation? How long did it go on? How often did it happen?

5. How many employees were involved in the violation? What were the levels of the employees involved in the violation? Was environmental compliance a criterion by which employees were judged? Were employees ever disciplined for environmental violations?

6. What efforts were made to remedy the results of the violation? What steps were taken to prevent a repetition of the violation?

7. What was the extent of good faith efforts to reach compliance agreements with government agencies? Were any such agreements fully carried out.

3.11. *Regulations Solution - 11.*

Term	Explanation
Liability	This means responsibility for an action. If an individual causes damage to property or other individuals, they are liable.
Strict Liability	This means responsibility without regard to negligence or care. A corporation could comply with all the applicable regulations in 1980 but when these regulations are changed in 1990, the corporation is liable for the new compliance.
Joint and Several Liability	In this case the responsibility is assigned (or shared) when several individuals (or corporations) do not perform properly, and it is not possible to divide the harm. If three plants contributed to a hazardous waste each one of them is liable to clean the site and mitigate the damages. This also includes the generators, the transporters, the storage facilities, and the operators. They are all collectively or individually responsible for damages.

3.11. *Regulations Solution - 11 (continued).*

Retroactive Liability	This is the case when a law is enacted, such as the Superfund Act; the liability goes back many years before the date of enactment.
Cradle-to-Grave Liability	This implies that the generator of the waste is responsible (liable) for the waste until its ultimate destruction. Simply selling the waste to another facility does not absolve liability.
Negligence	Negligence is an act or failure to act which breaches a responsibility that one person (or company) has to another person (or company) and which unintentionally results in harm to a person (or company) or to a person's (or company's) property. Negligent behavior leads to liability. If, for example, through a failure to exercise its responsibility to take proper care, a company allows a release of a toxic gas that kills the cows on a neighboring farm, then the company would be liable for the damages caused. Violation of a regulation would be virtual proof of negligence.
Trespass	Trespass to realty is the type of trespass most often used in environmental law cases. It involves the unlawful physical invasion of another's property that interferes with the use of that property. Trespass is independent of negligence. For example, even if there was not negligence, it is a trespass if gasoline from a gas station underground storage tank leaks out and flows under a neighbor's home, filling it with toxic and flammable vapors.
Nuisance	A nuisance is the use by a person of their property in a way that causes injury or annoyance to their neighbor. Allowing very bad smelling gases, dust, smoke or other annoying or harmful materials to drift over a neighbor's property would be examples of nuisance in environmental cases.

3.11. *Regulations Solution - 11 (continued).*

The understanding of these terms and liabilities implies that:
- it is very attractive to waste generators to dispose of the waste on-site under carefully controlled conditions
- it places a burden on the waste generator with the threat of future costs due to the improper action of others
- citizens are protected from the loss of property and/or health due to the action of others
- it is extremely important to select a reputable firm for waste management

3.12. *Regulations Solution - 12.*

1. - T
2. - T
3. - F (by SDWA)
4. - F (in CWA)
5. - F (by CWA)
6. - F (also applies to storer, treater, and transporter)
7. - F (not cleanup)
8. - F (not used much)
9. - T
10. - T

3.13. *Regulations Solution - 13.*

Treatment, storage and disposal of hazardous wastes require special permits, regardless of whether it is your waste or someone else's. This is true even if the waste is being safely and properly treated. As director of environmental affairs, you are required by law to inform the regulatory agencies of the infraction or face personal civil and criminal penalties.

3.14. *Regulations Solution - 14.*

Unfortunately, there are few easy, legal solutions to the problem, but following are a few that have been suggested:

1 - <u>Have the law changed</u> - The researcher who discovered the problem attempted to get the state environmental agency to change the rule, but failed after repeated attempts. One possible explanation for this is that environmental regulators are loathe to repeal regulations or to diminish their strength for fear of appearing to be controlled by industry. The "tougher is better" mentality of regulation still exists in most places.

2 - <u>Recycle the waste at a non-TSD facility</u> - Several individuals have pointed out that the regulators would have little chance of detecting shipments of the product to Texas for recycling if the companies did not file manifest documents on the material. The Texas recyclers would have no way of knowing that the product was technically a hazardous waste in Louisiana since it is identical to the product it receives from other generators in Texas. While there is a large economic incentive to do this ($1 to 2 million per year), the ethical and moral implications of circumventing state statutes are sufficient to prevent most people from doing this. In addition, this type of violation, if discovered and prosecuted, is a felony punishable by fine or prison or both.

3 - <u>Install on-site recycling facilities</u> - One possible solution is to reclaim the by-product on-site by use of an integrated processing facility. In some cases, this would be allowed without the expensive TSD permitting, but each case much be reviewed individually. The companies involved must determine whether an adequate rate of return can be realized from a relatively small reclamation facility that does not have the economies of scale realized by the Texas recyclers.

4 - <u>Move the process to another state</u> - It remains to be seen whether industry will threaten to close the plant and move to a more favorable regulatory environment in Texas or Ohio. If industry does employ this tactic and is successful in having the rules changed, the state regulators will definitely appear to have buckled under to industry.

3.15. *Regulations Solution - 15.*

The pollution prevention concerns contained in each of the Annexes of MARPOL can be summarized as follows:

ANNEX I - Regulations for the prevention of pollution by oil including: a list of oils, the International Oil Pollution Prevention (IOPP) Certificate, record keeping requirements, ship draught recommendations, and specifications for control of overboard discharges.

ANNEX II - Regulations for the control of pollution by Noxious Liquid Substances (NLS) in bulk including: guidelines for the categorization of Noxious Liquid Substances, lists of Noxious Liquid Substances, lists of other (non-polluting) liquid substances, record-keeping requirements, and standards and procedures for the discharge of Noxious Liquid Substances. Some of the procedures for the discharge of Noxious Liquid Substances that are given are: assessment of residues in cargo tanks, pumps and piping, prewash procedures, and ventillation procedures.

ANNEX III - Regulations for the prevention of pollution by harmful substances carried in packaged forms, freight containers, portable tanks or other forms.

ANNEX IV - Regulations for the prevention of pollution by sewage from ships.

ANNEX V - Regulations for the prevention of pollution by garbage from ships.

3.16. *Regulations Solution - 16.*

Discharge of oil from a tanker is prohibited by the International Convention for the Prevention of Pollution From Ships EXCEPT when ALL of the following conditions are met:

1. The tanker is not in a "special area."
2. The tanker is more than 50 nautical miles from the nearest land.
3. The tanker is en route.

3.16. *Regulations Solution - 16 (continued).*

4. The instantaneous rate of discharge of oil content does not exceed 30 L/nautical mi (NOTE - This condition was adopted in the 1992 amendment)
5. For existing tankers - the total quantity of oil discharged into the sea does not exceed 1/15,000 of the total quantity of the particular cargo of which the residue formed a part, and
   For new tankers - the total quantity of oil discharged into the sea does not exceed 1/30,000 of the total quantity of the particular cargo of which the residue formed a part
6. The tanker has in operation an oil discharge monitoring and control system and a slop tank arrangement as required by regulation.

3.17. *Regulations Solution - 17.*

The total coating volume for the Prime Coat per car is:
   Total prime coat per car = 6 L = 0.006 m^3

The total coating volume for the Top Coat per car is:
   Total top coat per car = 4 L = 0.004 m^3

The total mass of VOC generated per car is given as:
   Total VOC per car = VOC from prime coat/car + from top coat/car
   = (0.006 m^3)(300 kg/m^3) + (0.004 m^3)(520 kg/m^3)
   = 1.8 kg + 2.08 kg = 3.88 kg

The plant coats 1 car/min = 60 cars/h yielding a total hourly VOC emission rate of:
   Total VOC emission = (60 cars/h) (3.88 kg/car) = 232.8 kg/h

Thus, 232.8 kg/h VOC emissions is greater than the allowable emission rate of 80 kg/h. The plant, therefore, does not have a viable bubble.

Pollution prevention measures should be considered here, e.g., VOC substitution to reduce emission.

4.   Source   Reduction

4.1. *Source Reduction Solutions - 1*

a. Since the coating thickness and process throughput are to be unchanged, the required solids deposition rate can be computed first.

10 gal/h (0.25 gal solids/gal coating) (0.45 transfer efficiency)
= 1.125 gal solids/h

The current VOC emission rate can also computed for comparison with suggested modifications:

5.5 lb VOC/gal coating (10 gal/h) = 55 lb VOC/h

If the coating is reformulated to 35 vol % solids,

1.125 gal solids/h (1 gal/0.35 gal solids) (1/0.45 efficiency)
= 7.14 gal coating/h.

This 7.14 gal coating/h rate is a reduction of 28.6% from the original 10 gal/h usage rate.

In addition to less coating being used per hour, the new formula contains less VOCs since more of the volume is taken up by solids:

(0.65 gal VOC/gal coating) (7.36 lb VOC/gal VOC) (7.14 gal/h)
= 34.17 lb VOC/h

This represents a 38% reduction in VOC emissions as shown below:

(55 - 34.17 lb VOC/h)/(55 lb VOC/h) = 0.38

b. The process modification will result in less usage of coating since the new process is more efficient, thereby reducing VOC emissions. The required solids deposition rate is unchanged.

4.1. *Source Reduction Solutions - 1 (continued)*

1.125 gal solids/hr (1 gal coating/0.25 gal solids) (1/0.75 transfer efficiency) = 6 gal/h, a reduction of 40% from the original 10 gal/h usage rate.

6 gal/h (5.5 lb VOC/gal) = 33 lb VOC/h

This represents a 40% reduction in emissions from the original 55 lb VOC/hr rate as shown below:

(55 - 33 lb VOC/h)/(55 lb VOC/h) = 0.40

c. Using both modifications:

1.125 gal solids/h (1 gal coating/0.35 gal solids) (1/0.75 transfer efficiency) = 4.29 gal coating/h, a reduction of 57% from the original 10 gal/h rate.

0.65 gal VOC/gal coating (7.36 lb VOC/gal VOC) (4.29 gal/h) = 20.5 lb VOC/h

This represents a 63% reduction in emissions from the 55 lb VOC/h emission rate as:

(55 - 20.5 lb VOC/h)/(55 lb VOC/h) = 0.63

4.2. *Source Reduction Solutions - 2*

a. For the single-stage separation, a material balance on the metal residue yields:

$$f_{in}F + r_{in}R = f_1F + r_1R$$

There is no residue in the rinsewater feed, so $r_{in} = 0$. Rearrangement and substitution for r using the equilibrium condition, $r = f_1/\lambda$, yields:

4.2. *Source Reduction Solutions - 2 (continued)*

$$\frac{R}{F} = \frac{f_{in} - f_1}{r_1} = \lambda\left(\frac{f_{in} - f_1}{f_1}\right)$$

The fraction of residue removed, x, is defined as:

$$x = \frac{f_{in} - f_1}{f_{in}}$$

Substitution of x into the material balance and rearrangement yields:

$$\frac{R}{F} = \lambda\left(\frac{x}{1 - x}\right)$$

For 99% removal, x = 0.99, R = 99λF.

For a two-stage countercurrent operation,

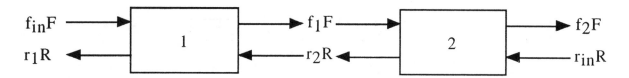

Material balances on the residue results in the following equations:

Stage 1: $f_{in}F + r_2R = f_1F + r_1R$

Stage 2: $f_1F + r_{in}R = f_2F + r_2R$

Again, there is no residue in the rinsewater feed, so $r_{in} = 0$.
Rearrangement and application of the equilibrium condition defining
r ($r_1 = f_1/\lambda$, $r_2 = f_2/\lambda$) results in the following two expressions:

Stage 1:

$$\frac{R}{F} = \lambda\left(\frac{f_{in} - f_1}{f_1 - f_2}\right)$$

4.2. *Source Reduction Solutions - 2 (continued)*

Stage 2:

$$\frac{R}{F} = \lambda \left(\frac{f_1 - f_2}{f_2} \right)$$

Substitution of x into the mass balance expressions, rearrangement and setting of the RHS of the Stage 1 and Stage 2 equations yields:

$$\frac{f_{in} - f_1}{f_1 - f_{in}(1 - x)} = \frac{f_1 - f_{in}(1 - x)}{f_{in}(1 - x)}$$

Rearrangement to obtain a quadratic in f_1 (with f_{in} replaced by f_o) produces:

$$f_1^2 - [(1 - x) f_{in}] f_1 - [x (1 - x) f_{in}^2] = 0$$

This equation is solved for f_1 to yield:

$$f_1 = \frac{f_{in} (1 - x)}{2} \left[1 \pm \sqrt{\frac{3x + 1}{1 - x}} \right]$$

Only the "+" term is physically reasonable, and when f_1 is substituted into the Stage 2 mass balance equation, the following expression results for a two stage countercurrent metal rinse tank:

$$\frac{R}{F} = \frac{x}{2} \left[\sqrt{\frac{3x + 1}{1 - x}} - 1 \right]$$

For 99% residue removal, (x = 0.99), the rinsewater requirements are:

For a single-stage unit:

$$R = \lambda \left(\frac{0.99}{1 - 0.99} \right) F = 99.0 \, \lambda F$$

For a two-stage countercurrent unit:

$$R = \frac{0.99 \, \lambda}{2} \left[\sqrt{\frac{3 (0.99) + 1}{(1 - 0.99)}} - 1 \right] = 9.5 \, \lambda F$$

4.2. *Source Reduction Solutions - 2 (continued)*

The rinsewater flowrate reduction is:

$$\frac{R \text{ (one stage)} - R \text{ (two stage)}}{R \text{ (one stage)}} = \frac{99 \lambda F - 9.5 \lambda F}{99 \lambda F}$$

$$= 0.90 \rightarrow 90\% \text{ reduction}$$

b. For fixed residue removal, the total mass of metals in the rinsewater will be the same; with reduced water duty, the metals concentration will increase. A smaller water volume now needs to be processed (or sewered). Alternatively, if the metals concentration is sufficiently high, it can be returned to the plating bath for reuse or can be recovered.

4.3. *Source Reduction Solutions - 3*

1. Always order the minimum amount of a chemical that you can foresee using and do not be tempted by the low price of bulk quantities. The presence of large quantities of unused chemicals takes up valuable space, increases the risk factor and always ends up more expensive when it is time for disposal. Increased risk factors include:

Storage of flammable liquids - 5 gal are more risky than 500 mL.

Decomposition of chemicals - over time the purity of a substance becomes suspect, particularly if it is susceptible to oxidation (aldehydes, ethers and alcohols).

Sampling opportunities increase with time - if the large stock bottle is frequently sampled, the risk of contamination increases.

Decomposition to form more dangerous compounds - diethyl ether (and other ethers) can form dangerous peroxides. Labels tend to decompose and fall off with time, increasing the danger of these end products.

4.3. *Source Reduction Solutions - 3 (continued)*

Obviously, chemical purity is more certain when you order a chemical as needed and not take it from one large bottle over a period of years. At least one chemical supplier has developed a line of chemicals for college lab experiments which provides exactly the amount needed as prescribed in the text. The cost is somewhat higher but there are savings in time, purity, storage and disposal.

2. The loss of labels on a bottle or can is simply a reflection of poor chemical inventory management. Purchase in smaller quantities and make certain that you adopt a "first in, first out" policy. To analyze a bottle of unknown material is expensive and time-consuming. For this reason, such unlabeled substances are much more expensive to dispose of through an outside agent since they charge for the analysis.

3. Whenever possible, substitute less hazardous chemicals and solvents for known hazards. There usually is no perfect solution, but tradeoffs in toxicity, carcinogenicity, flammability and other properties can usually be made. For instance, can ethyl acetate (flammable, non-toxic) be substituted for halogenated solvents? Can the procedure be modified to use less solvent? Take all reasonable precautions (ventilation, hoods, solvent covers, rubber gloves,etc.) to protect personnel and the environment.

4. Do not accept "free" chemicals from outside sources unless the chemicals are fresh, sealed, labeled, and on your shopping list. It is important to be certain that such gifts will be used within the foreseeable future since you wish to avoid costly disposal expenses at some later date. Also, old and/or partially used chemicals may be of questionable purity.

5. If done judiciously, waste acids and bases can be used to neutralize one another to produce a salt and water. In many cases, the salt is harmless and can enter the waste treatment plant. Other instances of treatment can be found in the small laboratory where hazardous substances can be rendered harmless by oxidation or reduction.

4.3. *Source Reduction Solutions - 3 (continued)*

In all cases and at all times, you should be aware of and be in compliance with all governing regulations.

4.4. *Source Reduction Solutions - 4.*

a. The balanced equation is:

$$3\ C_6H_{12}O + 2\ Na_2CrO_4 + 10\ H^+ \rightarrow 3\ C_6H_{10}O + 2\ NaH_2CrO_3 + 8\ H_2O$$
$$+ 2\ Na^+ + 2\ Cr^{+3} \qquad (1)$$

Refer to any general chemistry text for the details of balancing oxidation/reduction equations (See Problem 1.6). Briefly, however, the strategy is to separate the two processes, oxidation and reduction, into two half-reactions as if they occur independently.

reduction $\qquad CrO_4^{-2} + H^+ \rightarrow Cr^{+3} + H_2O$

oxidation $\qquad C_6H_{12}O \rightarrow C_6H_{10}O + H^+$

Each half-reaction is first balanced independently, using H_2O or H_3O^+ as needed to supply oxygen atoms (or hydrogen atoms) to either side.

$CrO_4^{-2} + 8H^+ + 3e^- \rightarrow Cr^{+3} + 4\ H2O$ for reduction, and

$C_6H_{12}O \rightarrow C_6H_{10}O + 2H^+ + 2e^-$ for oxidation

For an electrical balance, the reduction half-reaction is multiplied by a factor of 2 (6 electrons used). The oxidation half-reaction is multiplied by a factor of 3 (6 electrons produced). The two half-reactions are finally added to yield the solution given above.

The stoichiometry of the balanced reaction (Equation 1) shows that 2 lbmol of Na_2CrO_4 will produce 3 lbmol of product. One lbmol of product , $C_6H_{12}O$, has a mass of 98 lb. One thousand lb of product per day is equal to:

4.4. *Source Reduction Solutions - 4 (continued)*

$$1000 \text{ lb}/(98 \text{ lb/lbmol}) = 10.2 \text{ lbmol}$$

Using the ratio of 3:2 (product:reagent) from the balanced equation,

10.2 lbmol $C_6H_{12}O$ (2 lbmol Na_2CrO_4/3 lbmol $C_6H_{12}O$)
     = 6.8 lbmol of Na_2CrO_4 will be required

Since 1 lbmol of Na_2CrO_4 has a mass of 162 lb, then the mass required each day would be:

6.8 lbmol of Na_2CrO_4 (162 lb/lbmol Na_2CrO_4)
     = 1100 lb of Na_2CrO_4/d = 0.55 ton/d.

b. The chromium wastes are converted, by treatment with alkali, into chromium (III) hydroxide.

$$Cr^{+3} \rightarrow \text{alkali treatment} \rightarrow Cr(OH)_3$$

0.55 ton Na_2CrO_4/d generates 0.35 ton $Cr(OH)_3$

0.55 ton Na_2CrO_4/d (103 ton/gmol $Cr(OH)_3$)/(162 ton/gmol Na_2CrO_4) = 0.35 ton $Cr(OH)_3$/d

In 200 operating days, 70 ton of $Cr(OH)_3$ waste is generated. To calculate its volume, the density of 2.9 g/cm³ must be converted to ton/ft³. There are 454 g/lb, 2000 lb/ton and 1 ft/30.5 cm. Using these conversion factors, 1 g = 1.10 x 10^{-6} ton, and 1 cm³ = 3.52 x 10^{-5} ft³. Therefore 2.9 g/cm³ is:

2.9 g/cm³ (1.10 x 10^{-6} ton/g) (1 cm³/3.52 x 10^{-5} ft³) = 0.091 ton/ft³

This amount will occupy:

$$(70 \text{ ton } Cr(OH)_3)/(0.091 \text{ ton/ft}^3) = 770 \text{ ft}^3 \ Cr(OH)_3.$$

4.4. *Source Reduction Solutions - 4 (continued)*

c. Using the strategy outlined above, balancing the equation yields:

$$C_6H_{12}O + NaOCl \rightarrow C_6H_{10}O + NaCl + H_2O$$

From the solution in Part a above, 1000 lb product is equal to:

1000 lb (1 lbmol $C_6H_{12}O$ /98 lb $C_6H_{12}O$) = 10.2 lbmol
10.2 lbmol $C_6H_{12}O$ = 10.2 lbmol NaOCl

One lbmol of NaOCl weighs 74.4 lb, thus 10.2 lbmol weighs 760 lb. Household bleach is approximately 5.25% by weight NaOCl so the total weight required would be:

760 lb NaOCl/0.0525 = 14,400 lb NaOCl

One gal weighs approximately 8.0 lb so the total volume then would be:

14,400 lb NaOCl/(8 lb/gal NaOCl) = 1,809 gal of bleach/d

Since the plant operates 200 d/yr, the volume required is:

200 d (1,809 gal bleach/d) = 361,800 gal/yr.

4.5. *Source Reduction Solutions -5*

a. The basis for this calculation if an annual "throughput" of 1,000 samples. For the soxhlet extraction procedure, the following material balance diagram can be drawn:

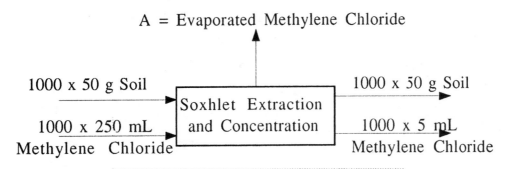

4.5. *Source Reduction Solutions - 5 (continued)*

For the SFE process, the materials balance diagram can be drawn as follows:

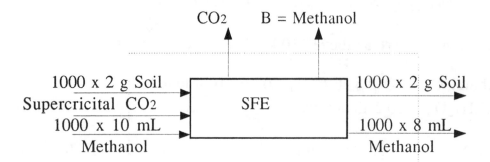

b. For the soxhlet extraction/concentration process, the material balance can be calculated as follows:

Solids input = 1,000 (50 g) = 50,000 g = 50 kg soil/yr
Methylene chloride input = 1,000 (250 mL) = 250,000 mL = 250 L/yr

Solids output = 1,000 (50 g) = 50,000 g = 50 kg soil/yr
Methylene chloride output = 1,000 (5 mL) + A = 5,000 + A mL = 5 + A L/yr
Input = Output

250 L/yr methylene chloride = 5 L/yr + A; A = 245 L/yr methylene chloride waste release

c. For the SFE process, the material balance can be calculated as follows:

The contaminated soil is all accounted for in the output stream, but is only 2 kg rather than 50 kg/yr produced from the soxhlet process.
The supercritical CO_2 is released from the system as a waste stream from the rear of the instrument, and is benign.
For the methanol:

Input = 1,000 (10 mL) = 10 L/yr methanol

4.5. *Source Reduction Solutions - 5 (continued)*

Output = 1,000 (8 mL) + B = 8,000 + B mL = 8 + B L/yr

Input = Output

10 L/yr = 8 + B L/yr; B = 2 L/yr waste methanol

d. The waste generated from the conventional process of soxhlet extraction versus that generated from the SFE process are summarized below:

Wastes	Soxhlet Extraction	SFE
Solids	50 kg/yr	2 kg/yr
Solvent	5 L/yr methylene chloride	8 L/yr methanol
Emission Losses	245 L/yr methylene chloride	2 L/yr methanol

From this table it is apparent that the soxhlet extraction process is of significantly more concern than the SFE process from both a health and safety, and an environmental perspective. Methylene chloride is a known human carcinogen, while methanol is relatively benign. Much higher quantities of methylene chloride are released untreated from the soxhlet process than is the methanol from the SFE process. If methylene chloride is to be removed/recovered during use of the soxhlet apparatus, its removal via a condenser unit would be difficult and costly. The methanol is exhausted from the SFE from a single port in the rear of the machine, and would be easy and relatively inexpensive to recover if desired.

4.6. *Source Reduction Solutions - 6*

a. The cost for soxhlet extraction and concentration for 1,000 soil samples/yr is calculated as follows:

4.6. *Source Reduction Solutions - 5 (continued)*

Chemical costs = 1,000 samples/yr (0.25 L/sample) ($10/L)
    = $2,500/yr
Disposal costs = 5 L/yr (1 gal/3.785 L) ($50/gal) = $66/yr
Glassware costs = $1,000/unit (1 unit/2 yr) = $500/yr
Labor = 1,000 samples/yr (1 d/8 samples) ($25/hr) (8 hr/d)
    = $25,000/yr

Total costs = Chemical + Disposal + Glassware + Labor = $2,500 + $66 +
    $500 + $25,000 = $28,066/yr = $28.07/sample

The cost for SFE for 1,000 soil samples/yr is calculated as follows:

Chemical costs = (1,000 samples/yr)(1 cylinder/150 sample)
    ($250/cylinder) = $1,667/yr
Disposal costs = (8 L/yr)(1 gal/3.785 L) ($10/gal) = $21/yr
Expendables costs = 1,000 samples/yr ($2/sample) = $2,000/yr
Equipment depreciation = $30,000/7 yr = $4,300/yr
Labor = 1,000 samples/yr (1 d/32 samples) ($25/hr) (8 hr/d)
    = $6,250/yr

Total costs = Chemical + Disposal + Expendables + Equipment
    Depreciation + Labor = $1,667 + $21 + $2,000 + $4,300 + $6,250
    = $14,238/yr = $14.24/sample

The cost advantage of the SFE over the soxhlet extraction process, not
    including reduced liability associated with CO_2 releases versus
    methylene chloride releases, is:

Cost savings/yr = $28,066 - $14,238 = $13,828/yr

The maximum payback period for the SFE is:

Payback = Equipment cost/Annual savings = $30,000/($13,828/yr)
    = 2.2 years

4.6. *Source Reduction Solutions - 6 (continued)*

b. Institutional resistance to this extraction method shift comes from the reliance of regulatory agencies on standard extraction/analytical methods for compliance, permitting and enforcement action activities. These savings may not be realized unless rigorous Quality Assurance/Quality Control (QA/QC) procedures are followed to validate the equivalence of these methods in a parallel performance study. Such comparisons should include: blank samples, spiked blanks, spiked inert solids (sand), environmental samples, duplicates, and spiked environmental samples. Once these studies are completed and they show equivalence, confidence in the SFE procedure should allow the switch over and initiation of this cost effective pollution prevention opportunity.

4.7. *Source Reduction Solutions - 7*

a. Annual product emission estimates can be made using the breathing loss and working loss equations presented in the problem statement

For the breathing losses, the input data needed are as follows:

M_v = 80 lb/lbmol @ 60 °F. This value must be corrected to the average ambient temperature of 40 °F. This can be done by reducing this value by the ratio of the vapor pressure of the mixture at 40 °F by that at 60 °F given in Table 1; (80 lb/lbmol) (0.8 psia)/(1.3 psia) = 49.2 lb/lbmol.

P = true vapor pressure at ambient conditions

The tank storage temperature = T_A + 3.5 from Table 2, therefore, the tanks storage temperature = 40 + 3.5 = 43.5 °F.

From Ts, P = (3.5/10) (0.2 psia) + 0.8 psia = 0.87 psia

P_a = 13.5 psia

D = 20 ft

H = 1/2 D = 10 ft since no information is provided about conical roof height

ΔT = Average maximum - Average minimum temperature = 60 °F - 30 °F = 30 °F

4.7. *Source Reduction Solutions - 7 (continued)*

Fp = paint factor = 1.33 from Table 3 assuming the Air Force maintains their tanks adequately

C = 0.9 from the problem statement

Kc = 1.0 from the problem statement

From these input data, the breathing loss from a tank is estimated to be:

L_B = 2.26 x 10^{-2} (49.2 lb/lbmol) [0.87 psia/(13.5 psia - 0.87 psia)]$^{0.68}$ (20 ft)$^{1.73}$ (10 ft)$^{0.51}$ (30 °F)$^{0.50}$ (1.33)(0.9)(1.0) = 681.4 lb/yr

The working losses are estimated from the equation given in the problem statement after first calculating the turnover rate of the product in the tank and the turnover factor, Kn.

$$N = (12{,}500 \text{ gal/d})(365 \text{ d/yr})/(50{,}000 \text{ gal}) = 91/\text{yr}$$

Kn @ N = 91 is interpolated from data given in the problem statement as:

$$Kn = 0.7 - [(91 - 50)/(100 - 50)] (0.7 - 0.47) = 0.51$$

L_w = 2.40 x 10^{-5} (49.2 lb/lbmol) (0.87 psia) (50,000 gal) (91/yr) (0.51)(1.0) = 2,384 lb/yr

Total losses = L_B + L_w = 681.4 lb/yr + 2,384 lb/yr = 3,065 lb/yr

Product release as a % of throughput
= [(3065 lb/yr)/6.4 lb/gal]/[(12,500 gal/d) (365 d/yr)]
= 0.0001 = 0.01%

b. Potential control techniques recommended by EPA include:

1. Installation of an internal floating roof and seal system with a "control" efficiency of 60 to 99%.
2. Installation of a vapor recovery system, i.e., vapor absorption, vapor compression, vapor condensation, or adsorption, with a control efficiency of 90 to 98%.

4.7. Source Reduction Solutions - 7 (continued)

3. Installation of a thermal oxidation system with a control efficiency of 96 to 99%.

These methods are very expensive and the later two may be incompatible with activity at an active fuel storage/transfer facility, particularly the thermal oxidation system.

One thing that you should notice from the EPA manual are the data presented in Table 3 which indicate that potentially significant reductions in breathing losses can be achieved with a simple change in tank color. This reduction can be quantified by repeating the calculations for a JP-4 tank using the data from the problem statement, assuming that the tank has been painted white to reduce breathing losses. These calculations are carried out by reestimating both the breathing and working losses as follows with a tank factor, $F_p = 1.00$, tank storage temperature $= T_a = 40\ °F$, and $P = 0.8$ psia:

$L_B = 2.26 \times 10^{-2}$ (49.2 lb/lbmol)[0.8 psia/(13.5 psia - 0.8 psia)]$^{0.68}$ (20 ft)$^{1.73}$ (10 ft)$^{0.51}$ (30 °F)$^{0.50}$ (1.0)(0.9)(1.0) = 482 lb/yr

$L_W = 2.40 \times 10^{-5}$ (49.2 lb/lbmol)(0.8 psia)(50,000 gal)(91/yr)(0.51) (1.0)
= 2,192 lb/yr

Total losses = $L_B + L_W$ = 482 lb/yr + 2,192 lb/yr = 2,674 lb/yr

Percent product loss reduction = [(3,065 lb/yr - 2,674 lb/yr)/(3,065 lb/yr)](100)
= 12.8%

Product release as a % of throughput
= [(2,674 lb/yr)/6.4 lb/gal]/[(12,500 gal/d) (365 d/yr)]
= 0.00009 = 0.009%

c. The low cost method of paint modification/replacement would be recommended over the more expensive/exotic approaches because of the low fraction of product loss from these release pathways. This color modification reduces overall emissions by nearly 13%, approximately 1 lb of product/tank/d, 61 gal/tank/yr, which

4.7. *Source Reduction Solutions - 7 (continued)*

amounts to a product cost savings of approximately \$40/tank/yr. The monetary return is low, but modification of tank color is also expected to be a no-cost item and can be implemented as part of the normal maintenance schedule for tank refurbishment that takes place on an on-going basis.

4.8. *Source Reduction Solutions - 8*

a. Losses from the fixed roof tank are estimated as follows:

Breathing loss:

$$\Delta T = 70°F - 50°F = 20°F$$

$$L_B = 2.26 \times 10^{-2} M_V \{P/(P_A - P)\} D^{1.73}H^{0.51}\Delta T^{0.5} F_P C K_P$$

$$= 2.26 \times 10^{-2} (62) \{7/(14.7 - 7)\} 80^{1.73} (40/2)^{0.51} 20^{0.5} (1)(1)(1)$$
$$= 51,463 \text{ lb/yr}$$

Working loss:

$$V = 3.141 (80^2)(40/4) \text{ ft}^3 (7.481 \text{ gal/ft}^3) = 1.5 \times 10^6 \text{ gal}$$
$$\text{Total throughput} = (10^6 \text{ bbl/yr})(42 \text{ gal/bbl}) = 4.2 \times 10^7 \text{ gal/yr}$$
$$N = (4.2 \times 10^7 \text{ gal/yr})/(1.5 \times 10^6 \text{ gal}) = 28/\text{yr}$$

$$L_W = 2.4 \times 10^{-5} M_v P V N K_n K_C$$

$$= 2.4 \times 10^{-5} (62)(7)(1.5 \times 10^6)(28)(1.0)(1.0) = 437,470 \text{ lb/yr}$$

Total loss, $L_T = L_B + L_W = 488,933 \text{ lb/yr}$

b. Losses from floating roof tank are estimated as follows:

Rim seal loss:

$$P^* = P/P_A /[1+(1 - P/P_A)^{0.5}]^2 = 7/14.7/[1 + (1 - 7/14.7)^{0.5}]^2 = 0.16$$

4.8. *Source Reduction Solutions - 8 (continued)*

$$L_R = K_S \, v^n \, P^* \, D \, M_v \, K_C$$
$$= 2.5(0.16)(80)(62)(1.0) = 1987.3 \text{ lb/yr}$$

Withdrawal loss:

$$L_W = 0.943 \, Q \, C \, W_L/D \, [1+(N_C \, F_C/D)]$$
$$= 0.943 \times 10^6 \, (0.0015)(5.6/80) \, [1+(1 \times 1/80)] = 100.4 \text{ lb/yr}$$

Deck fitting loss:

$$L_F = F_F \, P^* \, M_v \, K_C = 300 \, (0.16)(62)(1) = 2976 \text{ lb/yr}$$

Deck seal loss:

$$L_D = K_D \, S_D \, D^2 \, P^* \, M_v \, K_C = 0 \text{ lb/yr}$$

Total loss, $L_T = L_R + L_W + L_F + L_D = 5{,}060$ lb/yr

The total loss is considerably less in the case of the floating roof tank but it is still high enough to consider the installation of a vapor recovery system.

4.9. *Source Reduction Solutions - 9*

I_o = \$8,000,000, while CF = \$3,000,000/yr.

$$NPW = -8{,}000{,}000 + 3{,}000{,}000 \left[\frac{1}{1+r} + \frac{1}{1+r^2} + \frac{1}{1+r^3} + \frac{1}{1+r^4} \right]$$

$$= \$565{,}000$$

To obtain the DCF internal rate of return, NPW is set equal to 0, or:

4.9. *Source Reduction Solutions - 9(continued)*

$$0 = -8,000,000 + 3,000,000 \left[\frac{1}{1 + r} + \frac{1}{1 + r^2} + \frac{1}{1 + r^3} + \frac{1}{1 + r^4} \right]$$

or,

$$2.667 = \left[\frac{1}{1 + r} + \frac{1}{1 + r^2} + \frac{1}{1 + r^3} + \frac{1}{1 + r^4} \right]$$

The rate of return, r, is found by trial and error. The solution to this problem is found to be approximately 0.184, yielding a DCF rate of return = 18.4%. This means that if the interest rate is lower than 18.4%, the process modification is economically sound.

4.10. *Source Reduction Solutions - 10*

a. The environmental impact of laundered versus disposable diapers in the five categories presented in the problem statement is summarized as follows:

- net energy requirements - approximately the same
- atmospheric emissions - approximately the same
- industrial and postconsumer solid waste - disposal diapers generate approximately four times as much waste as the laundered diapers
- wastewater - cloth diapers produce approximately seven times more wastewater
- water volume requirements - cloth diapers require approximately four times more water volume than do the disposable diapers

Comments: The composition of the wastewater is not the same from laundered versus disposable diapers, and this may prove significant in certain areas, as may the significant difference in water requirements between the two types of diapers. It is not clear which product is superior, and will likely be determined on a situation specific basis.

4.11 *Source Reduction Solutions - 11*

a. Carrying out a material balance on the oil yields:

60 gal/yr = 0.30 M

Solving for for M; M = 200 gal/yr.

Carrying out a material balance on TCA results in:

1925 gal/yr = 0.7 (200 gal/yr) + E

Solving for E;

E = 1785 gal TCA/yr

b. 50% reduction in evaporative losses will save:

0.5 (1785 gal/yr) = 892.5 gal/yr

The associated cost savings is:

(892 gal/yr) (10^6 cm^3/264.17 gal) (1.34 g/cm^3) (1 lb/454 g)
($0.45/lb) = $4485/yr

c. 80% recovery of spent TCA represents the following cost savings:

0.8 [0.7 (200 gal/yr) (106 cm^3/264.17 gal) (1.34 g/cm^3)
(1 lb/454 g) ($0.45/lb)] = $563/yr

Reference: Higgins D. , M. May, M. Kostrzewa and H.W. Edwards. "Solvent Use Reduction in Microelectronics and Metal Fabrication Industries", *Proc. of the Conf. on Hazardous Waste Research*, Kansas State University, Manhattan, KS, May 29-30, 1991, pp. 537-551.

4.12. *Source Reduction Solutions - 12.*

It is not possible to simultaneously maximize a MSW recycling program for volume reduction, weight reduction, energy recovery and secondary material recovery. Following are the rank orders for the various scenarios defined by different recycling objectives:

Volume Reduction - This scenario is designed to maximize the reduction in the total volume of the waste stream by removing the component representing the largest volume percent first, with the removal of subsequent components based on their relative volume percents in the total waste stream. This rank order is as follows:

Table 1. Rank order for recycling of components of MSW based on maximizing volume reduction.

Component	Volume (%)
Paper	34.1
Other	21.6
Mixed Plastics	19.9
Yard Waste	10.3
Steel	9.8
Aluminum	2.3
Glass	2.0

Weight Reduction - This scenario is designed to maximize the reduction in the total weight of the waste stream by removing the component representing the largest weight percent first, with the removal of subsequent components based on their relative weight percents in the total waste stream. This rank order is as shown in Table 2.

Energy Recovery and Volume Reduction (No Incineration) - This scenario is designed to maximize non-thermal energy recovery (energy saved as a result of recycling, i.e., energy saved as less raw material requires processing when the finished material is reused), and the reduction in the total volume of the waste stream. This problem is solved by first determining the total energy equivalent of

4.12. *Source Reduction Solutions - 12 (continued)*

Table 2. Rank order for recycling of components of MSW based on maximizing weight reduction.

Component	Wt (%)
Paper	34.2
Other	21.6
Yard Waste	19.8
Mixed Plastics	9.2
Glass	7.1
Steel	7.0
Aluminum	1.1

each component in 1 ft^3 of MSW. The product of energy content and volume percent is then calculated for each component to give an indication of the potential for satisfying both constraints. The rank order for this problem is as follows:

Table 3. Rank order for recycling of components of MSW based on maximizing energy recovery and volume reduction with no incineration.

Component	Volume (%)	Volume (ft^3)	Energy Savings from Recycling[1] (Btu/ft^3)	Energy Recovery (Btu)	Energy x Vol
Paper	34.1	0.34	83,555	28,492	9716
Steel	9.8	0.10	30,277	2967	291
Aluminum	2.3	0.02	171,643	3948	91
Glass	2.0	0.02	57,289	1146	23
Other	21.6	0.22	nil	--	--
Mixed Plastics	19.9	0.20	nil	--	--
Yard Waste	10.3	0.10	nil	--	--

1 - Does not include post-consumer sorting or transportation costs.

Energy Recovery and Volume Reduction (with Incineration) - This scenario is designed to maximize thermal energy recovery, and the reduction in the total volume of the waste stream. This problem is

4.12. Source Reduction Solutions - 12 (continued)

solved by first determining the total energy equivalent of each component in 1 ft^3 of MSW, including combustion energy recovery for mixed plastics. The product of energy content and volume percent is then calculated for each component to give an indication of the potential for satisfying both constraints. The rank order for this problem is as follows:

Table 4. Rank order for recycling of components of MSW based on maximizing energy recovery and volume reduction with incineration of mixed plastics.

Component	Volume (%)	Volume (ft^3)	Energy Savings from Recycling[1] (Btu/ft^3)	Energy Recovery (Btu)	Energy x Vol
Paper	34.1	0.34	83,555	28,492	9716
Mixed Plastics	19.9	0.20	19,704	3921	780
Steel	9.8	0.10	30,277	2967	291
Aluminum	2.3	0.02	171,643	3948	91
Glass	2.0	0.02	57,289	1146	23
Other	21.6	0.22	nil	--	--
Yard Waste	10.3	0.10	nil	--	--

1 - Does not include post-consumer sorting or transportation costs.

Energy Recovery and Weight Reduction (No Incineration) - This scenario is designed to maximize non-thermal energy recovery and the reduction in the total weight of the waste stream. This problem is solved by first determining the total energy equivalent of each component in 1 lb of MSW, including combustion energy recovery for mixed plastics. The product of energy content and weight percent is then calculated for each component to give an indication of the potential for satisfying both constraints. The rank order for this problem is as shown in Table 5.

Energy Recovery and Weight Reduction (with Incineration) - This scenario is designed to maximize energy recovery, including thermal recovery from mixed plastics, and the reduction in the total weight

4.12. *Source Reduction Solutions - 12 (continued)*

Table 5. Rank order for recycling of components of MSW based on maximizing energy recovery with no incineration.

Component	Wt (%)	Wt (lb)	Energy Savings from Recycling[1] (Btu/lb)	Energy Recovery (Btu)	Energy x Wt
Paper	34.2	0.34	3008	1029	352
Steel	7.0	0.07	5450	382	27
Glass	7.1	0.07	2578	183	13
Aluminum	1.1	0.01	95,387	1049	12
Other	21.6	0.22	nil	--	--
Yard Waste	19.8	0.20	nil	--	--
Mixed Plastics	9.2	0.09	nil	--	--

1 - Does not include post-consumer sorting or transportation costs.

of the waste stream. This problem is solved by first determining the total energy equivalent of each component in 1 lb of MSW. The product of energy content and weight percent is then calculated for each component to give an indication of the potential for satisfying both constraints. The rank order for this problem is as follows:

Table 6. Rank order for recycling of components of MSW based on maximizing energy recovery and weight reduction with incineration of mixed plastics.

Component	Wt (%)	Wt (lb)	Energy Savings from Recycling[1] (Btu/lb)	Energy Recovery (Btu)	Energy x Wt
Paper	34.2	0.34	3008	1029	352
Mixed Plastics	9.2	0.09	14,000	1288	118
Steel	7.0	0.07	5450	382	27
Glass	7.1	0.07	2578	183	13
Aluminum	1.1	0.01	95,387	1049	12
Other	21.6	0.22	nil	--	--
Yard Waste	19.8	0.20	nil	--	--

1 - Does not include post-consumer sorting or transportation costs.

4.12. Source Reduction Solutions - 12 (continued)

Value of Recovered Materials on a Volume Basis - This scenario is designed to maximize the gross receipts for recovered materials from MSW on a unit volume basis. This problem is solved by determining the total revenue generated for each component in 1 ft^3 of MSW. The rank order for this problem is as follows:

Table 7.   Rank order for recycling of components of MSW based on maximizing revenue generation per ft^3 of waste.

Component	Volume (%)	Volume (ft^3)	Market Value ($/ft^3)	Total Revenue (Volume Basis ($))
Aluminum	2.3	0.02	0.468	1.08
Steel	9.8	0.10	0.056	0.54
Glass	2.0	0.02	0.167	0.33
Mixed Plastics	19.9	0.20	nil	--
Yard Waste	10.3	0.10	nil	--
Other	21.6	0.22	nil	--
Paper	34.1	0.34	-0.069	-2.37

Value of Recovered Materials on a Weight Basis - This scenario is designed to maximize the gross receipts for recovered materials from MSW on a unit weight basis. This problem is solved by determining the total revenue generated for each component in 1 lb of MSW. The rank order for this problem is as follows:

Table 8.   Rank order for recycling of components of MSW based on maximizing revenue generation per lb of waste.

Component	Wt (%)	Wt (lb)	Market Value ($/lb)	Total Revenue Wt Basis ($)
Aluminum	1.1	0.01	0.26	0.0029
Steel	7.0	0.07	0.01	0.0007
Glass	7.1	0.07	0.008	0.0005
Mixed Plastics	9.2	0.09	nil	--
Yard Waste	19.8	0.2	nil	--
Other	21.6	0.22	nil	--
Paper	34.2	0.34	-0.0025	-0.0009

4.12. *Source Reduction Solutions - 12 (continued)*

As can be seen from the results of these calculations, different recycling objectives will result in different waste components being targeted. It becomes imperative then to have a focused recycling program with built-in flexibility so that changes can be made in recycling activities to reflect changes occurring in recycling goals and changing recycling markets.

4.13. *Source Reduction Solutions - 13*

a. Since the molar ratio of N_2 to O_2 in air 79:21. The number of gmol of N_2 that correspond to one gmol of O_2 is 79/21 = 3.76 gmol.

b. For coal, from the stoichiometry of carbon oxidation, the total number of gmol in flue gas = CO_2 + 3.76 N_2 = 4.76 gmol. With 20% excess air, the stoichiometric equation takes the following form:

$$C + 1.2\ (O_2 + 3.76\ N_2) \rightarrow CO_2 + 0.2\ O_2 + 4.51\ N_2$$

and the volume of gas in the effluent = 5.71 gmol. The %CO_2 in flue gas = 17.5%

For natural gas, from its stoichiometry of combustion, the total number of moles in the flue gas = CO_2 + 2 H_2O + 7.52 N_2 = 10.52 gmol. With 20% excess air the stoichiometric equation takes the following form:

$$CH_4 + 1.2\ (2\ O_2 + 7.52\ N_2) \rightarrow CO_2 + 2\ H_2O + 0.4\ O_2 + 9.02\ N_2$$

and the volume of gas in the effluent = 12.42 gmol. The % CO_2 in flue gas = 8.0%.

c. One lb C releases 14,000 BTU and its conversion rate is 14,000 BTU/kW–h. Therefore, one lb C = 1 kW–h of energy. The C required/d for a 200 MW plant is therefore:

4.13. *Source Reduction Solutions - 13 (continued)*

Coal required = 200 MW/h (1000 kW/MW) (24 h/d) = 4,800,000 lb C/d

To determine the flowrate of the flue gas, convert lb C to gmol C as follows:

4,800,000 lb C/d (454 g/lb)/(12 g/gmol) = 181,600,000 gmol C/d

From Part b, 1 gmol C = 5.71 gmol flue gas, so the total molar flow rate is:

181,600,000 gmol C/d (5.71 gmol total gas/gmol C)
= 1,037,000,000 gmol total gas/d

Knowing 1 gmol = 22.4 L at STP, the total volumetric flow rate is:

1,037,000,000 gmol total gas/d (22.4 L/gmol)
= 23,200,000,000 L/d = 23,200,000,000 standard m^3/d

d. Capital and operating costs/d for CO_2 recovery are estimated as:

Capital and operating costs/d = $40,000,000/(10 yr x 365 d/yr)
= $10,959/d

The cost for recovering 1 gmol of CO_2 is (note 1 gmol C produces 1 gmol CO_2 during combustion):

($10,959/d)/(181,600,000 gmol CO_2/d) = $0.00006/gmol CO_2
= $0.06/kgmol CO_2 = 0.006 cents/gmol CO_2

Also, from Part c, 1 lb C = 454 g C = 37.83 gmol C = 1 kW–h. Therefore, the cost of adding a CO_2 recovery system is:

(37.83 gmol CO_2 /1 kW–h) (0.006 cents/gmol CO_2)
= 0.23 cents/kW–h = $0.0023/kW-h.

5. Recycling

5.1. *Recycling Solution-1.*

a. For a fresh ethane feed rate of 100,000 kg/h, the impact of one-pass ethane conversion efficiency on the plant recycle rate and hydraulic loading is determined through a material balance on the process. Material balance calculations were made for recycling flows of 40, 50, 60 and 70% conversion efficiency as described below.

Material balances are performed around each node, a point where streams enter and leave. The Total Flow material balance around each node is:

$$S_1 + S_4 - S_2 = 0 \qquad \text{[for the reactor]}$$

$$S_2 - S_4 - S_3 = 0 \qquad \text{[for the recovery section]}$$

Now the material balance is conducted for the components. There are two components in this problem, ethane (E) and products (P). For any stream, $E + P = S$.

$$E_1 + E_4 - E_2 = \text{ethane converted} \qquad \text{[for the reactor]}$$

$$E_2 - E_4 - E_3 = O \qquad \text{[for the recovery section]}$$

From the problem statement, it is also known that $S_1 = E_1 = S_3 = 100,000$ kg/hr and $S_4 = E_4$.

In addition, $E_2 = (1 - C/100)(100,000 + E_4)$ where C is the one-pass conversion efficiency.

Making these substitutions and rewriting the original four equations, the following result:

$$S_1 + S_4 - S_2 = 0 \qquad \text{(original)}$$
$$100,000 + E_4 - S_2 = 0 \qquad (1)$$

$$E_1 + E_4 - E_2 = \text{ethane converted} \qquad \text{(original)}$$
$$E_4 = E_2 + (\text{ethane converted} - E_1)$$
$$E_4 = E_2$$
$$E_4 = (1 - C/100)(100,000 + E_4) \qquad (2)$$

5.1. *Recycling Solution-1 (continued).*

$$E_2 - E_4 - E_3 = 0 \qquad (3)$$

For a given conversion efficiency, C, E_4 is calculated from Equation 2, then S_2 is determined from Equation 1, and finally E_3 is determined from Equation 3.

The solutions for 40, 50, 60 and 70% conversion efficiency are summarized in the Table below.

Table 1.   Summary of ethane and total flow rates for selected ethane conversion efficiencies.

Conversions	40%	50%	60%	70%
Stream Flow Rates (1000 kg/hr)				
E_1	100	100	100	100
E_2	150	100	66.7	42.9
E_3	0	0	0	0
E_4	150	100	66.7	42.9
S_1	100	100	100	100
S_2	250	200	166.7	142.9
S_3	100	100	100	100
S_4	150	100	66.7	42.9

P_1, P_2, P_3, P_4 are found by $P_n = S_n - E_n$

b. With the ethylene yield expression given in the problem statement, the net yield and ratio of total flow from the reactor, S_2, to net ethylene yield are calculated below as a function of ethylene conversion efficiency.

Table 2.   Summary of net ethylene yield and S_2/net ethylene yield ratio for selected ethane conversion efficiencies.

Conversions	40%	50%	60%	70%
Net ethylene yield (1000 kg/hr)	85.0	81.7	78.3	75.0
S_2/Net ethylene yield	2.94	2.45	2.13	1.91

5.1. *Recycling Solution-1 (continued).*

Although the lowest conversion produces the least raw material waste (P_2 - Net Yield of Ethylene), it also requires the highest flow rate of streams indicating higher utility costs and higher capital cost. These must be balanced as part of an economic analysis.

An alternative solution is possible by solving the material balance equations using matrix solution techniques. The solution matrix for the material balance equations written in Part a is as follows:

$$
\begin{bmatrix}
1 & 0 & 0 & 0 & 0 & 0 & 0 & 0 \\
0 & 1 & -1 & -1 & 0 & 0 & 0 & 0 \\
-1 & 0 & 1 & 0 & 0 & 0 & 0 & 0 \\
0 & 0 & 0 & 1 & 0 & 0 & 0 & 0 \\
-1 & 0 & 0 & 0 & 1 & 0 & 0 & 0 \\
0 & 0 & 0 & 0 & -g & 1 & 0 & -g \\
0 & 0 & 0 & 0 & 0 & -1 & 1 & 1 \\
0 & 0 & 0 & 0 & 0 & -1 & 0 & 1
\end{bmatrix}
\begin{bmatrix}
S_1 \\ S_2 \\ S_3 \\ S_4 \\ E_1 \\ E_2 \\ E_3 \\ E_4
\end{bmatrix}
=
\begin{bmatrix}
100,000 \\ 0 \\ 0 \\ 0 \\ 0 \\ 0 \\ 0 \\ 0
\end{bmatrix}
$$

$$\qquad\qquad \mathbf{A} \qquad\qquad\qquad \mathbf{b} \qquad\qquad \mathbf{c}$$

where $g = (1 - C/100)$. The matrix is inverted and solved to arrive at the results presented in Part a, as:

$$\mathbf{A} \times \mathbf{b} = \mathbf{c}$$
$$\mathbf{A}\ \mathbf{A^{-1}} \times \mathbf{b} = \mathbf{A^{-1}}\ \mathbf{c}$$
$$\mathbf{b} = \mathbf{A^{-1}}\ \mathbf{c}$$

5.2. *Recycling Solution-2.*

a. Cost avoidance is often the governing factor in the economics of pollution prevention. Consequently, in deciding whether to fund a recycling project, it is often easier to rank the alternatives in terms of lowest Total Present Worth costs. Total Present Worth calculations are carried out using the equations presented in the problem statement to this problem.

5.2. *Recycling Solution-2 (continued).*

In this problem, equal future costs for a period of n years are assumed. Applying the Biguns' internal rate of return requirement, the formula for the current situation with no investment for recycling becomes:

$$P = 0 + \sum_{n=1}^{5} \left[\frac{\$60,000}{(1 + 0.15)^n} \right]$$

Carrying out these calculations yields:

$$P = \$201,129$$

With a recycling investment, the Total Present Value is determined to be:

$$P = \$125,000 + \sum_{n=1}^{5} \left[\frac{\$25,000}{(1 + 0.15)^n} \right]$$

$$P = \$208,804 \text{ with recycling}$$

Therefore, when n = 5 years, $201,129 is less than the cost of the system with recycling of $208,804, and it does not make strict economic sense to invest the capital in recycling.

b. With n = 1,...,10 for Part b, calculations yield

$$P = \$301,126$$

with no recycling. With recycling an n = 1,...,10,

$$P = \$250,469 \text{ with recycling}$$

Therefore, when n = 10 years, $301,126 is more than the cost of the system with recycling of $250,469, and it does make strict economic sense to invest the capital in recycling, thereby incurring lower total cost.

c. The rate of return is obtained when the present worth of the savings from recycling equals the initial investment, i.e., solving for i when:

5.2. *Recycling Solution-2 (continued).*

$$P \text{ w/recycling} = \$125,000 + \Sigma \left[\frac{\$25,000}{(1 + i)^n} \right] \text{ equals}$$

$$P \text{ w/ out recycling} = \Sigma \left[\frac{\$60,000}{(1 + i)^n} \right]$$

Therefore, i must be solved for in the following equation:

$$\$125,000 + \Sigma \left[\frac{\$25,000}{(1 + i)^n} \right] = \Sigma \left[\frac{\$60,000}{(1 + i)^n} \right]$$

Using Newton's method, let

$$g = \$125,000 + \Sigma \left[\frac{\$25,000}{(1 + i)^n} \right] - \Sigma \left[\frac{\$60,000}{(1 + i)^n} \right]$$

then

$$g' = -n \Sigma \left[\frac{\$25,000}{(1 + i)^{n+1}} \right] + n \Sigma \left[\frac{\$60,000}{(1 + i)^{n+1}} \right]$$

The recursion formula for Newton's method is:

$$i = i_{guess} - \frac{g (i_{guess})}{g' (i_{guess})}$$

After five iterations, i = 24.99%.

Therefore, the recycling rate of return exceeds the company's required internal rate or return when the duration of the project is given as 10 years.

d. Calculating the Present Worth of the cost of each of the options using the cost data presented in the problem statement is a simple matter of taking the sum of the cost data for each. This results in the following Total Present Worth for each option.

5.2. *Recycling Solution-2 (continued).*

Year	Costs, $ No Recycling	Costs, $ With Recycling
1	64,321	22,492 + 125,000
2	65,356	23,451
3	68,879	25,012
4	71,034	27,914
5	72,308	29,037
Total Present Worth	$227,202	$209,133

Since the recycling case has the lower cost on a present Total Present Worth basis, the recycling should be done.

5.3. *Recycling Solution-3.*

a. The following schematic represents the KNO_3 crystallization process as described in the problem statement:

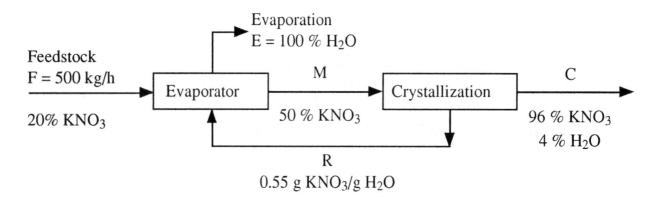

b. Material balances for the entire system are as follows:

For the total mass into and out of the unit:

$F = 5,000$ kg/h $= C + E$

For KNO_3, the material balance is expressed as:

0.2 F kg/h $= 0.96$ C kg/h; $F = 0.96/0.2$ C $= 4.8$ C kg/h

Substitution of F from the system balance into the KNO_3 balance yields:

5.3. *Recycling Solution-3 (continued).*

$$5{,}000 \text{ kg/h} = 4.8 \text{ C}; \quad C = 1041.7 \text{ kg/h}$$

From the system balance, the value of E is:

$$5{,}000 \text{ kg/h} = 1041.7 \text{ kg/h} + E; \quad E = 3958.3 \text{ kg/h}.$$

Material balances around the crystallization unit are as follows:

For the total mass into and out of the unit:

$$M = 1041.7 \text{ kg/h} + R$$

For KNO_3, the material balance is expressed as:

$$0.5 \text{ M} = 0.96 \,(1041.7 \text{ kg/h}) + (0.55 \text{ g } KNO_3 \,/1.55 \text{ g total}) \text{ R}$$
$$0.5 \text{ M} = 1000 \text{ kg/h } KNO_3 + 0.355 \text{ R}; \quad M = 2000 \text{ kg/h } KNO_3 + 0.71 \text{ R}$$

Equating M from both the total system and KNO_3 material balances yields:

$$1041.7 \text{ kg/h} + R = 2000 \text{ kg/h } KNO_3 + 0.71 \text{ R}$$
$$0.29 \text{ R} = 958.3 \text{ kg/h}; \quad R = 3304.5 \text{ kg/h}$$

An alternative solution can be given by describing the material balance equations in matrix format. The material balance equations are shown in simplified form below:

$$F = 500$$
$$F + R = M + E \qquad \text{(Overall balance)}$$
$$0.2 \text{ F} + 0.55/1.55 \text{ R} = 0.5 \text{ M} + 0 \text{ E} \qquad \text{(One component)}$$

$$M = C + R \qquad \text{(Overall balance)}$$
$$0.5 \text{ M} = 0.96 \text{ C} + 0.55/1.55 \text{ R} \qquad \text{(One component)}$$

In matrix form these equations can be written as:

$$
\begin{bmatrix}
1 & 0 & 0 & 0 & 0 \\
1 & 1 & -1 & -1 & 0 \\
0 & -1 & 1 & 0 & -1 \\
0.2 & 0.355 & -0.5 & 0 & 0 \\
0 & -0.355 & 0.5 & 0 & -0.96
\end{bmatrix}
\begin{bmatrix}
F \\ R \\ M \\ E \\ C
\end{bmatrix}
=
\begin{bmatrix}
5{,}000 \\ 0 \\ 0 \\ 0 \\ 0
\end{bmatrix}
$$

$$\qquad\qquad\text{A} \qquad\qquad\qquad\qquad \text{b} \qquad\quad \text{c}$$

5.3. *Recycling Solution-3 (continued).*

Inverting the matrix and solving for **b** yields the following results from **b** = **A** $^{-1}$ **c**;

F = 5,000; R = 3304.6; M = 4346.26; E = 3958.33; C = 1041.67

c. To do the repetitive calculations required to develop a curve, the material balance was organized into matrix form as shown in Part b, and solved by inverting the matrix using a TRUE BASIC™ program, the source code of which is listed below.

```
DIM  a(5,5),a1(5,5),b(5,1),c(5,1)
MAT READ a
DATA  1,0,0,0,0
DATA  1,1,-1,-1,0
DATA  0,-1,1,0,-1
DATA  0.2,0.355,-0.5,0,0
DATA  0,-0.355,0.5,0,-0.96
LET  c(1,1)=5000
MAT  ai=inv(a)
MAT  b=ai*c
SET  zonewidth(10)
PRINT "   F"," R","  M"," E"," C"
PRINT  b(1,1),b(2,1),b(3,1),b(4,1),b(5,1)
GET KEY z
CLEAR
SET  window  0.2,1.1,-1000,11000
BOX  LINES  0.3,1,0,10000
FOR  i=1 to 10
   BOX  LINES  0.3,1,0,1000*i
NEXT i
FOR  i=1 to 7
   BOX  LINES  0.3,0.3+0.1*i,0,10000
NEXT i
FOR  conc=0.3 to 0.9 step 0.001
   LET  a(4,3)=-conc
   LET  a(5,3)=conc
   MAT  ai=inv(a)
   MAT  b=ai*c
   PLOT  conc,b(2,1)
NEXT conc
END
```

5.3. *Recycling Solution-3 (continued).*

In the program, the concentration of M is varied from 0.3 to 0.9 (g KNO_3/g H_2O) in 0.001 increments, the matrix is inverted and the results are plotted as shown in the figure below.

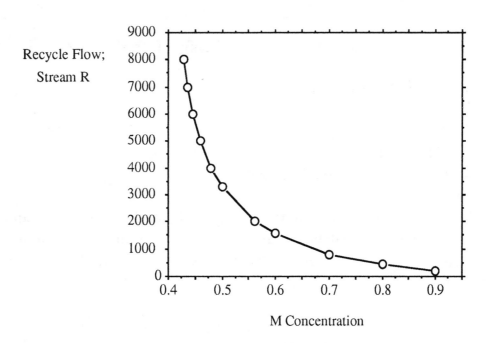

The main conclusion from this analysis is that if the concentration in stream M gets much lower in KNO_3 than originally specified, the necessary recycle rate will become unreasonably high.

5.4. *Recycling Solution-4.*

a. Mass balances for the degreasing process without the vapor recovery system are as follows:

Unit 1: $F_{13} + F_{12} = F_{01} + F_{21}$ (1)
Unit 2: $F_{21} + F_{24} = F_{12}$ (2)

The following additional information is available:

$$F_{01} = 1 \text{ kg (basis)} \qquad (3)$$
$$F_{13} = 0.6 (F_{01} + F_{21}) \qquad (4)$$
$$F_{21} = 0.8 F_{12} \qquad (5)$$

Five equations may be used to solve for the five material streams.

5.4. *Recycling Solution-4 (continued).*

From Equations 1, 3, and 4:

$$F_{12} = F_{11} + F_{21} - F_{13} = 1 + F_{21} - 0.6 (1 + F_{21}) = 0.4 (1 + F_{21})$$

or, $1.25 \ F_{21} = 0.4 (1 + F_{21})$

From the above equation, $F_{21} = 0.471$ kg/kg of fresh solvent.

Using Equations 2 through 6,

$F_{12} = 0.583$ kg/kg of fresh solvent.
$F_{13} = 0.882$ kg/kg of fresh solvent.
$F_{24} = 0.118$ kg/kg of fresh solvent.

b. To do the repetitive calculations required to develop the requested
   curve, the material balance was organized into matrix form as shown
   below, and solved by inverting the matrix using a TRUE BASIC™
   program.

$$F_{12} = F_{24} + F_{21}$$
$$F_{01} + F_{21} = F_{13} + F_{12}$$
$$F_{13} = 0.6 (F_{01} + F_{21})$$
$$F_{21} = 0.8 \ F_{12}$$
$$F_{01} = 1.0 \ kg$$

In matrix form these equations can be written as:

$$\begin{bmatrix} 0 & 1 & 0 & -1 & -1 \\ 1 & -1 & -1 & - & 0 \\ -0.6 & 0 & 1 & -0.6 & 0 \\ 0 & -0.8 & 0 & 1 & 0 \\ 1 & 0 & 0 & 0 & 0 \end{bmatrix} \begin{bmatrix} F_{01} \\ F_{12} \\ F_{13} \\ F_{21} \\ F_{24} \end{bmatrix} = \begin{bmatrix} 0 \\ 0 \\ 0 \\ 0 \\ 1 \end{bmatrix}$$

$$\mathbf{A} \qquad\qquad \mathbf{b} \qquad \mathbf{c}$$

Inverting the matrix and solving for **b** yields the following results from
b = **A**$^{-1}$ **c** with the emission factor = 0.6:

$F_{01} = 1.0; \ F_{12} = 0.588; \ F_{13} = 0.88; \ F_{21} = 0.47; \ F_{24} = 0.12$

5.4. *Recycling Solution-4 (continued).*

The TRUE BASIC™ source code for the solution of this problem is listed below.

```
DIM  a(5,5),a1(5,5),b(5,1),c(5,1)
MAT READ a
DATA  0,1,0,-1,-1
DATA  1,-1,-1,1,0
DATA  -0.6,0,1,-0.6,0
DATA  0,-0.8,0,1,0
DATA  1,0,0,0,0
LET  c(5,1)=1
MAT  ai=inv(a)
MAT b=ai*c
SET  zonewidth(10)
PRINT "    F01","    F12","    F13","    F21","    F24"
PRINT  b(1,1),b(2,1),b(3,1),b(4,1),b(5,1)
GET KEY z
CLEAR
SET  window  0,1.3,-0.400,2.400
BOX  LINES 0.3,1,0,2.000
FOR  i=1 to 10
    BOX  LINES 0.3,1,0,0.200*i
NEXT i
FOR  i=1 to 7
    BOX  LINES 0.3,0.3+0.1*i,0,2.000
NEXT i
FOR  conc=0.3 to 0.9 step  0.002
    LET  a(3,1)=-conc
    LET  a(3,4)=-conc
    MAT  ai=inv(a)
    MAT  b=ai*c
    PLOT  conc,b(2,1)
NEXT conc
END
```

In the program, the solvent emitted from the degreaser (the emission factor) is varied from 0.3 to 0.9 in 0.002 increments, the matrix is inverted and the results are plotted as shown in the figure below.

5.4. *Recycling Solution-4 (continued).*

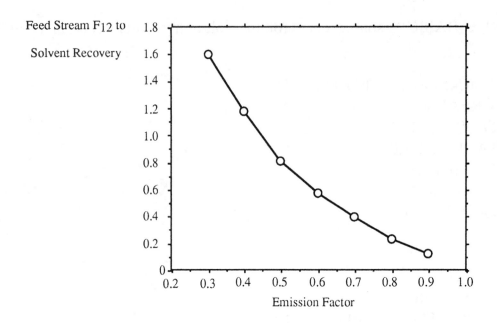

The main conclusion from this analysis is that if the emitted solvent is high, then the solvent recovery unit will be underloaded. This would also suggest a need for a vapor recovery system.

5.5. *Recycling Solution-5.*

a. Material balances for the degreasing process with a vapor recovery system are as follows:

Unit 1: $F_{01} + F_{21} + F_{31} = F_{13} + F_{12}$ (1)
Unit 2: $F_{12} = F_{21} + F_{24}$ (2)
Unit 3: $F_{13} = F_{31} + F_{35}$ (3)

The following additional information is available:

$$F_{01} = 1 \text{ kg (basis)} \qquad (4)$$
$$F_{13} = 0.6 \, (F_{01} + F_{21} + F_{31}) \qquad (5)$$
$$F_{21} = 0.8 \, F_{12} \qquad (6)$$
$$F_{31} = 0.9 \, F_{13} \qquad (7)$$

Equations 1 through 7 may be solved to yield seven unknowns. The results are:

5.5. *Recycling Solution-5 (continued).*

$$F_{12} = 2.857 \text{ kg/kg of fresh solvent}$$
$$F_{13} = 4.286 \text{ kg/kg of fresh solvent}$$
$$F_{21} = 2.286 \text{ kg/kg of fresh solvent}$$
$$F_{24} = 0.571 \text{ kg/kg of fresh solvent}$$
$$F_{31} = 3.857 \text{ kg/kg of fresh solvent}$$
$$F_{35} = 0.429 \text{ kg/kg of fresh solvent}$$

b. To do the repetitive calculations required to develop the requested curve, the material balance equations, Equations 1 through 7, were organized into matrix form as shown below, and solved repeatedly by inverting the matrix:

$$
\begin{bmatrix}
1 & 0 & 0 & 0 & 0 & 0 & 0 \\
1 & -1 & -1 & 1 & 0 & 1 & 0 \\
-0.6 & 0 & 1 & -0.6 & 0 & -0.6 & 0 \\
0 & 1 & 0 & -1 & -1 & 0 & 0 \\
0 & -0.8 & 0 & 1 & 0 & 0 & 0 \\
0 & 0 & 1 & 0 & 0 & -1 & -1 \\
0 & 0 & -0.9 & 0 & 0 & 1 & 0
\end{bmatrix}
\begin{bmatrix}
F_{01} \\
F_{12} \\
F_{13} \\
F_{21} \\
F_{24} \\
F_{31} \\
F_{35}
\end{bmatrix}
=
\begin{bmatrix}
1 \\
0 \\
0 \\
0 \\
0 \\
0 \\
0
\end{bmatrix}
$$

$$\qquad\qquad\quad \mathbf{A} \qquad\qquad\qquad\qquad \mathbf{b} \qquad \mathbf{c}$$

Inverting the matrix and solving for $\mathbf{b}$ yields the results shown in Part a for each process stream from $\mathbf{b} = \mathbf{A}^{-1}\mathbf{c}$ with an emission factor = 0.60.

Using the following TRUE BASIC™ program, the solvent emitted from the degreaser (the emission factor) is varied from 0.3 to 0.9 in 0.002 increments, the matrix is inverted and the results are plotted as shown in the figure below.

The main conclusion from this analysis is that if the emitted solvent is high, then the solvent recovery unit loading decreases. However, with the addition of the vapor recovery unit, the loading to the solvent recovery unit varies by less than a factor of two. This is far improved over the system without vapor recovery, where the loadings to the solvent recovery system varied by an order of magnitude.

5.5. *Recycling Solution-5 (continued).*

The TRUE BASIC™ source code for the solution of this problem is listed below.

```
DIM  a(7,7),a1(7,7),b(7,1),c(7,1)
MAT READ a
DATA  1,0,0,0,0,0.0
DATA  1,-1,-1,1,0,1,0
DATA  -0.6,0,1,-0.6,0,-0.6,0
DATA  0,1,0,-1,-1,0,0
DATA  0,-0.8,0,1,0,0,0
DATA  0,0,1,0,0,-1,-1
DATA  0,0,-0.9,0,0,1,0
LET  c(1,1)=1
MAT  ai=inv(a)
MAT  b=ai*c
SET  zonewidth(11)
PRINT "    F01","   F12","   F13","   F21"
PRINT  b(1,1),b(2,1),b(3,1),b(4,1)
PRINT
PRINT
PRINT "   F24","   F31","   F35"
PRINT  b(5,1),b(6,1),b(7,1)
GET KEY z
CLEAR
SET  window  0,1.3,-1.00,6.00
BOX  LINES  0.3,1,0,5.000
FOR  i=1 to 10
   BOX  LINES  0.3,1,0,0.500*i
NEXT i
FOR  i=1 to 7
   BOX  LINES  0.3,0.3+0.1*i,0,5.000
NEXT i
FOR  conc=0.3 to 0.9 step 0.002
   LET  a(3,1)=-conc
   LET  a(3,4)=-conc
   MAT  ai=inv(a)
   MAT  b=ai*c
   PLOT  conc,b(2,1)
NEXT conc
END
```

5.5. *Recycling Solution-5 (continued).*

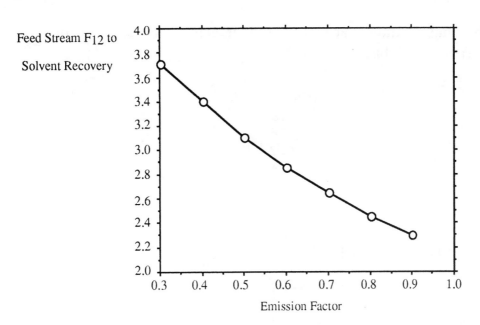

Feed Stream F_{12} to Solvent Recovery (y-axis)

Emission Factor (x-axis)

c. In order to compare the two processes it is desirable to do the calculation on the basis of the amount of solvent disposed off with the sludge. This amount is related to the amount of material treated. For comparison, the amount of fresh solvent used and the amount of solvent lost to the atmosphere is calculated for a system without vapor recovery (Case 1, Problem 5.4 solution) versus a system with vapor recovery (Case 2, Problem 5.5 solution) on the basis of a unit amount of solvent disposed off with the sludge.

Case 1: $F_{01}/F_{24} = 1/0.118 = 8.5$ kg fresh solvent/kg waste solvent
   $F_{13} (F_{01}/F_{24}) = 0.882/0.118 = 7.5$ kg lost/kg waste solvent

Case 2: $F_{01}/F_{24} = 1/0.571 = 1.75$ kg fresh solvent/kg waste solvent
   $F_{35} (F_{01}/F_{24}) = 0.429/0.571 = 0.75$ kg lost/kg waste stream

By employing the vapor recovery system, the amount of solvent lost due to evaporation, and fresh solvent requirements, are reduced considerably.

5.6. *Recycling Solution-6.*

a. The mass balance around the degreasing tank is as follows using a 1 lb fresh PCE basis:

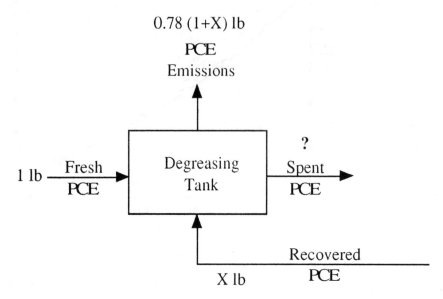

Input = Fresh + Recycled PCE = 1 lb + X lb

Output = PCE Emissions + Spent PCE = 0.78 (1 + X) + Spent PCE
Spent PCE = (1 - 0.78) (1 + X) = 0.22 (1 + X) lb PCE

b. The mass balance around the solvent recovery unit is as follows:

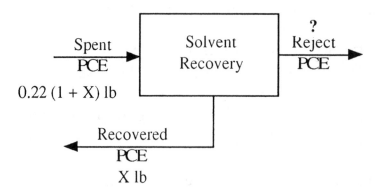

Input = Spent PCE = 0.22 (1 + X) lb

Output = Recycle PCE + Reject PCE; Recycle PCE = 75% of Spent PCE
Output = 0.75 (0.22 (1 + X)) + Reject PCE

Input = Output
    0.22 (1 + X) = 0.75 (0.22 (1 + X)) + Reject PCE
    Reject PCE = (1 - 0.75) (0.22 (1 + X)) = 0.055 (1 + X)
        = 0.055 + 0.055 X lb

5.6. *Recycling Solution-6 (continued).*

c. The mass balance around the entire system is as follows:

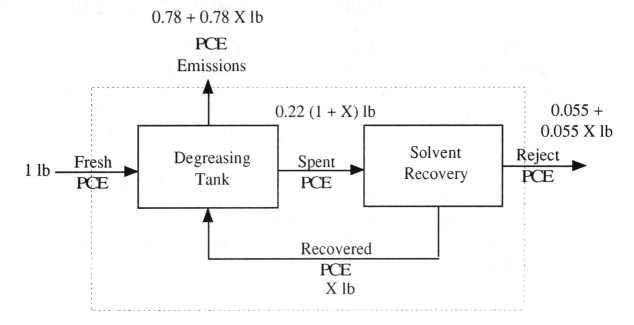

Input = 1 lb PCE

Output = PCE Emissions + Spent PCE
   = (0.78 + 0.78 X) + (0.055 + 0.055 X)

Input = Output
   1 lb PCE = (0.78 + 0.78 X) + (0.055 + 0.055 X)
   1 lb PCE = 0.835 + 0.835 X
   X = 0.165/0.835 = 0.198 = 0.20 lb PCE

d. The lb PCE emitted/lb fresh PCE utilized from the mass balance calculations is:

PCE Emissions = 0.78 + 0.78 X = 0.78 + 0.78 (0.20)
   = 0.936 = 0.94 lb PCE Emitted/lb fresh PCE

e. If the emission factor was lower, the flow rates to the solvent recovery unit and the recycle stream would be higher. Additionally, there would be less PCE lost from the system. To determine the effect of the emission factor on system flow streams, the equations above were solved using three different emission factors, 0.78, 0.60, and 0.40. These results are summarized below.

5.6. *Recycling Solution-6 (continued).*

<u>Emission Factors</u>

	<u>0.78</u>	<u>0.60</u>	<u>0.40</u>
Fresh PCE	1.0	1.0	1.0
Recovered PCE	0.198	0.429	0.818
Spent PCE	0.263	0.571	1.091
PCE Emissions	0.934	0.857	0.727
Reject PCE	0.066	0.142	0.273

The sum of the Recovered and Fresh PCE provides a measure of the degreasing capability of the system/kg feed. Notice that as the emission factors go down, this sum goes up significantly, providing more degreasing capacity/kg fresh feed as degreasing tank emissions are reduced.

5.7. *Recycling Solution 7.*

a. The total energy requirements and emission values for the no-recycle case for each product are as follows:

Product 1: energy = 185 J/bag + 464 J/bag = 649 J/bag
Product 2: energy = 724 J/bag + 905 J/bag = 1629 J/bag
Product 1: emissions = 0.414 g/bag + 1 g/bag = 1.414 g/bag
Product 2: emissions = 1.463 g/bag + 1.478 g/bag = 2.914 g/bag

b. The total energy requirements and emissions for the 25% recycle case are determined based on a material balance description for the systems as shown below:

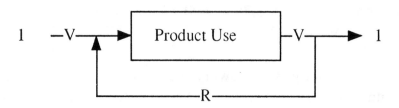

$R + 1 = V$
$x = \text{Recycle Rate} = R/V$

therefore, $R = x\,V$, $V = R/x$, and $1 + xV = V$.

5.7. *Recycling Solution-7 (continued)*.

Finally, the product use rate, V, is:

$$V = \frac{1}{(1 - x)}$$

Note that E1 and G1 are not impacted by product recycle for a given amount of fresh feed (fb = fresh bag):

Product 1: energy = 185 J/fb + 464/(1 - 0.25) J/fb = 804 J/fb
Product 2: energy = 724 J/fb + 905/(1 - 0.25) J/fb = 1931 J/fb
Product 1: emissions = 0.414 g/fb + 1/(1 - 0.25) J/fb = 1.747 g/fb
Product 2: emissions = 1.463 g/bag + 1.478/(1 - 0.25) J/fb
                     = 3.434 g/bag

For x = 0.25, V = 1.333 bu/fb (bu-bags utilized), therefore, the numbers given above must be divided by 1.333 to yield values on a per bu basis:

Product	Energy, J/bu	Emissions, g/bu
1	603	1.310
2	1449	2.576

c. The total energy requirements and emissions for the 50% recycle case are calculated as follows:

Product 1: energy = 185 J/fb + 464/(1 - 0.5) J/fb = 1113 J/fb
Product 2: energy = 724 J/fb + 905/(1 - 0.5) J/fb = 2534 J/fb
Product 1: emissions = 0.414 g/fb + 1/(1 - 0.5) J/fb = 2.414 g/fb
Product 2: emissions = 1.463 g/bag + 1.478/(1 - 0.5) J/fb
                     = 4.419 g/bag

For x = 0.5, V = 2 bu/fb, therefore, the numbers given above must be divided by 2 to yield values on a per bu basis:

Product	Energy, J/bu	Emissions, g/bu
1	556	1.207
2	1267	2.210

d. The total energy requirements for both products are equated, with the 2:1 Product 1:Product 2 requirement as follows:

5.7. *Recycling Solution-7 (continued)*.

$$2 \ [185 + 464/(1-x)] = 724 + 905/(1-x)$$

Solving for x yields a required recycle rate of 93.5% for the energy requirements of the two products to be equivalent. Based on this result, substitution of Product 2 for Product 1, even with the 1:2 equivalence would not be practical as it is unlikely that 93.5% of Product 2 could be effectively recycled.

5.8. *Recycling Solution 8.*

a. An estimated bulk density is determined for each waste component based on the ratio of its weight% to volume% in MSW. This approximate bulk density indicates which material will give the highest weight reduction per unit of volume reduction. The results of this calculation are shown below:

Component	Wt%/Vol%
Paper/Paperboard	1.0
Glass	3.6
Aluminum	0.5
Recyclable Steel	0.7
Plastics	0.5
Yard Waste (compostable)	1.9
All Other (non-recyclable)	1.0

Of the recyclable material, glass and paper have the highest approximate density. Therefore, as much glass as possible should be recycled, with the bulk of the 25% volume reduction coming from the recycling of paper.

Based on the volume of glass in MSW, a minimum of 23% volume reduction must come from Paper/Paperboard. This means that a minimum of 67% of the Paper/Paperboard in the waste must be recycled to meet the 25% volume reduction goal with recycling it and glass alone.

b. The value of recyclable material in a ton of MSW is calculated below based on the data presented in the problem statement. Based on these calculations, Paper/Paperboard has the highest value in raw

5.8. *Recycling Solution-8 (continued).*

MSW. Recall from Part a that Paper/Paperboard is also the highest wt%/vol% component. Therefore, all volume reduction should come from recycling paper to maximize the monetary benefit from recycling efforts.

Component	ton/ton MSW	$/ton MSW
Paper/Paperboard	0.341	8.53
Mixed Glass	0.071	2.84
(If Separated)		(7.10)
Aluminum	0.011	0.33
Recyclable Steel	0.070	7.00
Mixed Plastics	0.092	4.6
(If Separated)		(18.4)
Yard Waste (compostable)	0.198	0
All Other (non-recyclable)	0.216	0

The value of waste paper varies widely in some regions of the U.S., and has become nearly worthless from time to time as supply exceeds the recycling capacity of reprocessing mills.

Additionally, if plastics and glass are separated, their value goes up significantly. If separation is to take place, then plastics should be recycled as much as possible, followed by steel and sorted glass, particularly if the waste paper market is not strong in a given area.

5.9. *Recycling Solution-9.*

a. Some glass products contain metal, paint, organics, and other contaminants that render them unsuitable for food and/or beverage containers. There is also the additional problem of sterilizing the containers before use in food and beverage packaging, and some glass products may not be amenable to current sterilization techniques. Glass items that should not be recycled and the reasons why are summarized below.

Item	Reason
Light bulbs	Contain filiments, other metals
Dishes	Contain paint
Windshields	Composite containing metal and organics
Plate glass	May contain metals, organics
Ceramic glass	Contains inorganic salts and metals

5.9. *Recycling Solution-9 (continued).*

   b. Yes. Most plastics contain plasticizers, fillers and solvents. Some plastics also contain residual monomers. Plastic beverage containers contain metals and organic compounds in the inks and dyes used for their design.

5.10. *Recycling Solution-10.*

   a. First determine the total number of gmol of air in the room. Keep in mind that for an ideal gas 1 gmol occupies 22.4 L (or 0.0224 m^3) at STP.

   The room temperature is at 22°C = 295 K
   Total air gmol in room
       = (1100 m^3) (1000 gmol/22.4 STP m^3) (273 K/295 K)
       = 45,440 gmol air

   Also from the ideal gas law, PV = nRT

       (1 atm)(1100 m^3)(1000 L/m^3) = n (0.08221 atm-L/gmol-K)(295 K)
       n = (1,100,000 atm-m^3)/(24.2 atm-L/gmol) = 45,420 gmol air

   The mole fraction of HC = 1.5 gmol HC/45,440 gmol air
       = 0.000033 gmol of HC/gmol of air = 33 ppm
       = 33,000 ppb >> 850 ppb.

   Yes there is a health risk.

   b. To decrease the environmental hazard associated with the reactor, several things can be done:

      i. If allowed by local ordinance, the reactor may be contained in a sealed room and vapors emitted if the reactor ruptured may be discharged into a hood or a duct to vent the reactor to the atmosphere. This is not the preferred solution.
      ii. The laboratory could install an improved air circulation system that continually provided replacement of the laboratory air to limit the maximum hydrocarbon concentration to below the 850 ppb health limit. The net outflow of air should be monitored and treated if necessary by scrubbing or adsorption before being released into the atmosphere.

5.10. *Recycling Solution-10 (continued).*

c. To improve pollution prevention, input substitution may be possible, i.e., it may be possible to replace the material in the reactor with material that would have a lower vapor pressure, and/or be less of a health hazard. In addition, the reactor seals may be modified or eliminated if possible. Double seals would be less likely to leak, improving the safety and reduce the pollution hazard of the existing reactor.

6. Treatment/Disposal

6.1. *Treatment/Disposal Solution-1.*

The major design considerations that must be addressed include the following:

1. Type and volume of hazardous and non-hazardous wastes to be landfilled.
2. Life expectancy of the landfill during its active operating period.
3. Topography and soil characteristics at the site and its vicinity.
4. Climatic conditions throughout the year.
5. Surface water and ground water in the vicinity of the landfill.
6. Collection and treatment of surface runoff.
7. Soil cover requirements for individual containment cells.
8. Anticipated quality and quantity of leachate.
9. Selection of leachate collection and treatment systems.
10. Monitoring of ground water and surface water during operation and beyond.
11. Selection of venting systems for gaseous products.
12. Selection of flexible membrane and other impermeable liners
13. Closure and post-closure plans.
14. Alternative uses during the post-closure period.
15. Impact on human health as well as on animals and vegetation.

6.2. *Treatment/Disposal Solution-2.*

For solid waste transportation, the following equation can be used to predict the potential liability of waste hauling:

$$PL = \frac{\$10,000}{\text{ton spilled}} \left(\frac{x \text{ tons spilled}}{\text{overturn}}\right) \left(\frac{u \text{ tons waste}}{\text{yr}}\right) \left(\frac{1 \text{ trip}}{z \text{ tons}}\right) \left(\frac{1 \text{ spill}}{4000 \text{ trips}}\right)$$

where PL = potential liability, x = estimated tons spilled during an accident, u = total tons of waste disposed of/yr, and z = typical amount of waste loaded/truck at the plant.

For this problem, the following substitutions can be made:

$$PL = \frac{(\$10,000/\text{ton}) (2 \text{ tons/accident}) (32,000 \text{ tons disposed/yr})}{(4000 \text{ trips/accident}) (4 \text{ tons/trip})}$$

6.2. *Treatment/Disposal Solution-2 (continued).*

PL = $40,000/yr, or on a $/ton basis:

$$\frac{\$40,000/yr}{32,000 \ tons/yr} = \$1.25/ton$$

As this problem demonstrates, pollution prevention activities which
reduce the generation of waste can substantially reduce the cost of
future liabilities, as well as required disposal costs.

6.3. *Treatment/Disposal Solution-3.*

Using a basis of 100 gmol of flue gas, the molar composition of the
mixture and gmol of C and O are as follows:

	moles	C	O_2
CO_2	7.5	7.5	7.5
CO	1.3	1.3	0.65
O_2	8.1		8.1
N_2	83.1		
TOTAL	100	8.8	16.25

Conducting an O_2 balance assuming combustion using atmospheric air
with an O_2/N_2 ratio of 21/79 yields a total of:

83.1 gmol N_2 (21/79) = 22.1 gmol O_2

This O_2 is associated with the nitrogen in the gas. The missing oxygen is
in the form of H_2O and amounts to:

O_2 in combustion air - O_2 in water-free flue gas = O_2 in H_2O

22.1 gmol O_2 - 16.25 gmol O_2 = 5.85 gmol O_2 in H_2O, or
(5.85 gmol O_2 in H_2O) (2 gmol H_2O/gmol O_2) = 11.7 gmol H_2O

The gmol of H combusted to form this H_2O is:

11.7 gmol H_2O (2 gmol H/gmol H_2O) = 23.4 gmol H

This hydrocarbon therefore has a H/C gmol ratio of 23.4 gmol H/8.8
gmol C:

6.3. *Treatment/Disposal Solution-3 (continued).*

$$\frac{23.4}{8.8} = 2.66 \frac{H}{C} \text{ atoms}$$

The lowest integer ratio corresponding to 2.66 is 8/3. Although the ratios 16/6, 31/12, etc., also satisfy the H/C ratio value of 2.66, there are no known hydrocarbons that fit these ratios. For example, the general formula for an alkane is C_nH_{2n+2}; all other hydrocarbons will have fewer hydrogens/carbon atom. If C = 6, H_{max} = 14, if C = 12, H_{max} = 26, etc., thus, the hydrocarbon in the gas stream is the alkane, C_3H_8, propane.

6.4. *Treatment/Disposal Solution-4.*

This is a mass balance problem. The mass of chromium disposed will be the amount that is present in the dewatered sludge. First estimate the amount of dewatered sludge from the given data.

The amount of filtrate/d = filter area (volume of filtrate/h-ft^2) (h of operation/d)
= 150 ft^2 (10 gal/h-ft^2) (16 h/d) = 24,000 gal/d

Develop mass balance equations relating the raw sludge and the dewatered sludge as follows:

Mass Balance Equation 1:

Mass of raw sludge = mass of dewatered sludge + mass of filtrate

Mass Balance Equation 2:

Mass of solids in the raw sludge = mass of solids in the dewatered sludge

With these equations, the following input data are available:

mass of raw sludge not known
mass of dewatered sludge not known
mass of filtrate = 24,000 gal/d (8.34 lb/gal) = 200,160 lb/d
mass of solids in the sludge = 0.05 (mass of raw sludge)

6.4. *Treatment/Disposal Solution-4 (continued).*

Therefore, from a mass balance on the solids:

$$0.05 \text{ (raw sludge)} = 0.48 \text{ (dewatered sludge)}$$

and from a mass balance on the water:

$$0.95 \text{ (raw sludge)} = \text{Filtrate} + 0.52 \text{ (dewatered sludge)}$$

From Equation 1, dewatered sludge = 0.05 (raw sludge)/0.48. Substituting this into Equation 2 yields:

$$0.95 \text{ (raw sludge)} = \text{Filtrate} + 0.52 \text{ [0.05 (raw sludge/0.48)]}$$

or,

$$0.95 \text{ (raw sludge)} = 200,160 \text{ lb/d} + 0.52 \text{ [0.05 (raw sludge/0.48)]}$$

Solving for the mass of raw sludge in the equation above yields:

$$\text{mass of raw sludge} = (200,160 \text{ lb/d})/(0.8958) = 223,434 \text{ lb/d}$$

The mass of dewatered sludge is found as:

$$\text{mass of dewatered sludge} = 0.05 \text{ (223,434 lb/d)}/0.48 = 23,274 \text{ lb/d}$$

The mass of chromium disposed of/d is:

$$\text{chromium disposed of/d} = 120 \text{ mg/kg (23,274 lb/d) (0.4536 kg/lb)}$$
$$= 1,266,872 \text{ mg/d} = 1.267 \text{ kg/d} = 2.79 \text{ lb/d}$$

The chromium disposed/yr = 2.79 lb/d (7 d/wk) (50 wk/yr)
   = 977.6 = 978 lb/yr

6.5. *Treatment/Disposal Solution-5.*

a. The principle involved is simply the conservation of mass. There must be the same amount of material, and the same number of atoms on both sides of the equation. Often oxidation reactions will be quite complex and require a good deal of thought and strategy to

6.5. *Treatment/Disposal Solution-5 (continued).*

balance. The equation in the problem statement, however, can be treated as if the oxygen atom is simply moved from the chlorine atom to the sulfur atom and no consideration of the various oxidation states is necessary. Keep in mind that the ammonium ion, NH_4^+, and the sodium ion, Na^+, are "spectator ions" and are not involved in the oxidation reaction. The equation could be more simply written as:

$$OCl^- + S^{2-} \rightarrow Cl^- + SO_4^{2-}$$

This is balanced as a standard chemical reaction, requiring four hypochlorite ions for each sulfide: the balanced form is as follows:

$$8\ OCl^- + 2\ S^{2-} \rightarrow 8\ Cl^- + 2\ SO_4^{2-}$$

b. The strategy is to determine the number of gmol of each reactant in their respective containers and then to use the stoichiometry of the equation to determine how many gal of bleach are required to react completely with the $(NH_4)_2S$.

i. Calculate the mass of $(NH_4)_2S$, without water, in each container.

2.3 kg (0.201) = 0.46 kg = 460 g

Convert 0.46 kg to gmol of $(NH_4)_2S$. $(NH_4)_2S$ weighs 68 g/gmol. (This is obtained by adding up the atomic weights of the various elements.)

460 g /(68 g/gmol) = 6.8 gmol of $(NH_4)_2S$

ii. Calculate the number of gmol of NaOCl required for complete reaction. From the equation, it can be seen that 1 gmol of the sulfide requires 4 gmol of the hypochlorite. Therefore, 6.8 gmol of the sulfide will require four times as many moles, or 27.2 gmol, of NaOCl.

iii. Determine the number of moles of NaOCl in a one gallon container of bleach.

6.5. *Treatment/Disposal Solution-5 (continued).*

8.7 lb of bleach contains 5.25% NaOCl, or 0.46 lb of NaOCl. This is equivalent to 207 g.

Convert g of NaOCl to gmol of NaOCl. One gmol of NaOCl has a mass of 74.5 g.

$$(207 \text{ g/gal})/(74.5 \text{ g/gmol}) = 2.8 \text{ gmol/gal}$$

iv. Calculate the number of gal of bleach required and the total cost of the bleach.

From Part ii, 27.2 gmol of NaOCl are required, while 1 gal = 2.8 gmol

$$27.2 \text{ gmol}/(2.8 \text{ gmol/gal}) = 9.7 \text{ gal of household bleach}$$

At $0.89/gal of bleach, the cost would be $8.63/gal of sulfide treated, not considering the labor costs.

The total cost of treatment of a lab pack is:

$$12 \times \$8.63 = \$103.56/\text{lab pack}$$

c. The cost for disposal by a reputable outside contractor would be $350 for 12 containers of $(NH_4)_2S$ or $29.17 for each container. On this basis alone, it would be more economical to dispose of the chemical by oxidation with household bleach.

This exercise does not take into account several factors: the cost of one's labor, the necessity for a large (> 10 gal) vat with cooling and a large, well-ventilated space. One should also purchase a reference manual which outlines the proper procedure and necessary precautions.

Note: The calculations in this exercise can also be carried out on a lbmol basis if desired. The strategy for solution is the same.

6.6. *Treatment/Disposal Solution-8.*

a. The degradation reaction for a generic straight-chain aliphatic hydrocarbon is:

$$C_nH_{2n+2} + (3n+1)/2 \ O_2 \rightarrow n \ CO_2 + (n+1) \ H_2O$$

The molecular weight of the hydrocarbon is:

(12n + 2n + 2)= 14n + 2 g/gmol,

requiring:

(3/2) 32n + (1/2) 32 = 48n + 16 g O_2/gmol hydrocarbons degraded.

Therefore, (48n + 16)/(14n + 2) mass units of O_2 are needed per unit mass of hydrocarbon.

Because of their low vapor pressure, hydrocarbons with n > 4 carbons are likely to be involved in biological reactions as they remain in the soil and do not volatilize completely before biodegradation begins. The mass ratio as a function of n is then:

Carbon Number n	O_2/hydrocarbon mass ratio = (48n + 16)/(14n + 2)
5	3.55
10	3.49
15	3.47
∞	3.43

b. Hydrogen peroxide releases oxygen according to the following stoichiometric relationship:

$$2 \ H_2O_2 \rightarrow 2 \ H_2O + O_2$$

Two gmol of hydrogen peroxide are required per gmol of oxygen generated. Therefore,

2 (34) = 68 kg hydrogen peroxide are required per 32 kg oxygen generated.

The released gasoline mass = 5,000 gal (3.78 L/gal) (0.001 m³/L) (700 kg/m³) = 13,230 kg

6.6. *Treatment/Disposal Solution-6 (continued).*

The required O_2 mass = 3.5 (13,230 kg) = 46,305 kg

The required H_2O_2 mass then = 68/32 (46,305 kg) = 98,398 kg

The cost for the chemicals alone then is

$2.20/kg (98,398 kg)   = $ 216,476!!!

6.7. *Treatment/Disposal Solution-7.*

The volume of liquid resulting from the absorption of ammonia from a contaminated gas stream is calculated as follows for a 100 gmol gaseous mixture (see the figure below). This calculation is based on the assumption that no water evaporates from the absorber.

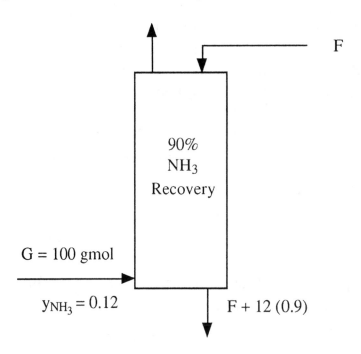

For the inlet gas stream:

G = 100 gmol
G_{NH_3} = 12 gmol
G_{N_2} = 88 gmol

Since 90% of the ammonia is absorbed, (0.9) (12) represents the gmol of ammonia in the liquid stream leaving the absorber.

6.7. *Treatment/Disposal Solution-7 (continued).*

For these conditions, the mole fraction of ammonia in the liquid leaving the absorber, x, is:

$$x = \frac{(0.9)(12)}{F + (0.9)(12)}$$

The mass fraction of ammonia, w, that is specified to be 0.04, is then given by:

$$w = \frac{(0.9)(12)(17)}{(F)(18) + (12)(17)} = 0.04$$

where 17 and 18 represent the molecular weights of ammonia and water, respectively. Solving this equation for F yields:

F = 244.8 gmol water, and
W = 4406 g water.

The mass of water is added to the mass of ammonia to obtain the total mass of the aqueous solution:

A = 4406 g water + 12 (0.9) (17) g ammonia
A = 4590 g aqueous solution

The density of the solution = 0.98 g/cm^3, so that the total volume of solution produced after treatment of 100 gmol of the gaseous mixture is:

4590 g aqueous solution/(0.98 g/cm^3) = 4684 cm^3
= 4.7 L/100 gmol gaseous mixture.

6.8. *Treatment/Disposal Solution-8.*

a. The number of transfer units are calculated as:

NTU = Z/HTU = 4 m/1 m = 4

The effluent water concentration is calculated by solving the NTU equation given in the problem statement as:

6.8. *Treatment/Disposal Solution-8 (continued).*

$$4 = \frac{5}{5 - 1} \ln \frac{4 \, (C_{in}/C_{out}) + 1}{5}$$

$$\frac{C_{in}}{C_{out}} = 30.4$$

C_{out} = 30 ppm/30.4 = 0.99 = 1 ppm

$$\text{Removal} = \frac{C_{in} - C_{out}}{C_{in}} = \left[\frac{30 \text{ ppm} - 1 \text{ ppm}}{30 \text{ ppm}} \right]$$

$$= 97\%$$

b. The air-water ratio is calculated from the stripping factor equation given in the problem statement as:

$$\frac{G}{L} = R \frac{P_t}{H} = 5 \frac{1 \text{ atm}}{321 \text{ atm}} = 0.0154$$

$$\frac{G''}{L''} = 1325 \, (0.0154) = 20.4$$

The tower area is $\pi \, (1.5 \text{ m})^2/4 = 1.77 \text{ m}^2$; hence, the air loading is

$$G'' = (3 \text{ m}^3/\text{min})/(1.77 \text{ m}^2) = 1.7 \text{ m}^3/\text{min-m}^2$$

The water loading rate is then:

$$L'' = (1.7 \text{ m}^3/\text{min-m}^2)/20.4 = 0.083 \text{ m}^3/\text{min-m}^2$$

The maximum water flow is:

$$(0.083 \text{ m}^3/\text{min-m}^2)(1.77 \text{ m}^2)$$
$$= (0.147 \text{ m}^3/\text{min}) \, (264 \text{ gal/m}^3) = 39 \text{ gpm}$$

6.9. *Treatment/Disposal Solution-9.*

The reaction which produces Fe_2O_3 is:

$$4 \, Fe^{2+} + 3 \, O_2 + 8 \, e^- \rightarrow 2 \, Fe_2O$$

6.9. *Treatment/Disposal Solution-9 (continued).*

The molecular weights of each component of the reaction are as follows:

Fe = 55 g/gmol, O_2 = 32 g/gmol, and
Fe_2O_3 = 2 (55) + 3 (16) = 158 g/gmol

Hence, 220 kg of Fe^{2+} produce 316 kg of Fe_2O_3

The Fe_2O_3 produced from the oxidation of the iron entering the tower is calculated as 316/220 times the mass of Fe^{2+} entering the tower or:

Fe^{2+} = (35 mg Fe^{2+}/L) (10^{-6} kg/mg) (30 gpm) (3.78 L/gal) (1440 min/d)
= 5.72 kg Fe^{2+}/d entering the tower, and

Fe_2O_3 = (316 kg Fe_2O_3/220 Fe^{2+}) (5.72 kg Fe^{2+}/d) = 8.21 kg Fe_2O_3/d

Since 20% of the Fe_2O_3 sludge is flushed out, the amount of Fe_2O_3 accumulated in the stripper is 0.8 (8.21 kg/d) = 6.57 kg/d.

As the sludge clogs the pore space, the lost volume in the stripper will be:

$$(6.57 \ kg/d)/(3200 \ kg/m^3) = 2.05 \ x \ 10^{-3} \ m^3/d$$

Since the volume of packing in the tower is:

0.56 (4 m) (π) [(0.6 m)^2/4] = 0.63 m^3

a 10% pore space reduction is equal to 0.063 m^3. This reduction will take place after:

(0.063 m^3)/(2.05 x 10^{-3} m^3/d) = 31.5 = 32 d.

Hence, the stripper should be cleaned with an HCl solution every 32 days. HCl is suggested as the cleaning solution since the $FeCl_2$ which forms is soluble and can be flushed from the tower.

6.10. *Treatment/Disposal Solution-10.*

a. Calculate $[Na^+]$ from initial conditions. Using a charge balance at a pH = 12, the $[Na^+]$ is found as:

$$[H^+] + [Na^+] = [OH^-]$$

$$10^{-12} \text{ M} + [Na^+] = 10^{-2} \text{ M}$$

$$[Na^+] = 10^{-2} \text{ M}$$

After neutralization, at pH = 7, there is negligible CO_3^{2-} as indicated from the dissociation constant k_{A2}:

$$k_{A2} = 5.61 \times 10^{-11} \text{ M} = \frac{(10^{-7} \text{ M}) [CO_3^{-2}]}{[HCO_3^-]}$$

$$\frac{[CO_3^{-2}]}{[HCO_3^-]} = 5.61 \times 10^{-4}$$

Thus, the overall charge balance can be solved for $[HCO_3^-]$ as:

$$[H^+] + [Na^+] = [OH^-] + [HCO_3^-] + 2 [CO_3^{-2}]$$

$$10^{-7} \text{ M} + 10^{-2} \text{ M} = 10^{-7} \text{ M} + [HCO_3^-]$$

$$[HCO_3^-] = 10^{-2} \text{ M}$$

The carbonic acid species concentration can be estimated using the expression for k_{A1} and substitution for $[H^+]$ and $[HCO_3^-]$ as:

$$k_{A1} = 4.30 \times 10^{-7} \text{ M} = \frac{[H^+] [HCO_3^-]}{[H_2CO_3]}$$

$$k_{A1} = 4.30 \times 10^{-7} \text{ M} = \frac{(10^{-7} \text{ M}) (10^{-2} \text{ M})}{[H_2CO_3]}$$

$$[HCO_3^-] = 2.33 \times 10^{-3} \text{ M}$$

Summing the carbon species, yields:

6.10. *Treatment/Disposal Solution-10 (continued).*

$$\Sigma C = [H_2CO_3] + [HCO_3^-] = 2.33 \times 10^{-3} \text{ M} + 10^{-2} \text{ M} = 1.23 \times 10^{-2} \text{ M}$$

The CO_2 mass per bbl of beer produced is calculated as:

$$\left(\frac{1.23 \times 10^{-2} \text{ gmol CO}_2}{\text{LH}_2\text{O}}\right)\left(\frac{44 \text{ g CO}_2}{\text{gmol CO}_2}\right)\left(\frac{119 \text{ LH}_2\text{O}}{\text{bbl H}_2\text{O}}\right)\left(\frac{1 \text{ lb}}{454 \text{ g}}\right)\left(\frac{5 \text{ bbl H}_2\text{O}}{\text{bbl beer}}\right)$$

$$= 0.71 \frac{\text{lb CO}_2}{\text{bbl beer}} \text{ @ 100\% absorption}$$

With process efficiency at 50% absorption, the CO_2 requirement then is 1.42 lb CO_2/bbl beer.

b. The fraction of CO_2 used for rinse tank neutralization at a production rate of 9 lb CO_2/bbl beer is:

$$\left(\frac{1.42 \text{ lb CO}_2/\text{bbl beer}}{9 \text{ lb CO}_2/\text{bbl beer}}\right) = 0.16 = 16\%$$

c. The CO_2 must be vented from the fermentor and subsequently stored until needed for rinsewater neutralization.

6.11. *Treatment/Disposal Solution-11.*

a. The number of pore volumes for flushing is dependent on the total TCE volume, the total TCE mass, the solubility of TCE in water, and the volume of water in a pore volume.

initial TCE volume = $(0.35)(0.20) 1 \text{ m}^3 = 0.07 \text{ m}^3$
initial TCE mass = $(0.07 \text{ m}^3) (1,470 \text{ kg/m}^3) = 103 \text{ kg}$

The water volume needed to dissolve 103 kg of TCE is:

$$V = (103 \text{ kg}) (1000 \text{ g/kg})/1.1 \text{ g/L} = 93.6 \text{ m}^3$$

6.11. *Treatment/Disposal Solution-11 (continued).*

The water in 1 pore volume is:

$0.35 \ (1 \ m^3) \ (0.8) = 0.28 \ m^3$

The number of pore volumes to flush out the TCE is:

$93.6 \ m^3/0.28 \ m^3 = 334.3$

b. Removal time of TCE is related to the permeability of the aquifer and the number of pore volumes that must be moved.

Flow through one face of the soil cube is:

$Q = (0.02 \ m/d) \ (1 \ m^2) = 0.02 \ m^3/d$

The time to dissolve all of the residual TCE is:

$t = 93.6 \ m^3/(0.02 \ m^3/d) = 4,680 \ d = 12.8 \ yr$

This required time can also be determined from the effective pore velocity and average retention time of the ground water in one pore volume.

pore velocity $= (0.02 \ m/d)/[(0.35) \ (0.80)] = 0.071 \ m/d$

The time to travel through the cube, i.e., the pore volume retention time is:

pore retention time $= 1 \ m/(0.071 \ m/d) = 14 \ d$

Therefore, the time to move 334.3 pore volumes is:

Time to dissolve all TCE $= 334.3$ pore volumes (14 d/pore volume) $= 4680 \ d = 12.8 \ yr.$

NOTE: As the TCE is flushed out, the residual saturations will change so that the water volume in the soil at the beginning of the process is less than the volume at the end. Since the water residual saturation is high, this correction is small, and can safely be neglected. Also, this assumes equilibrium conditions during "pumping;" in reality this never occurs, with reported extraction efficiency generally varying from only 10 to 50%. As this problem illustrates, it takes a long time to remediate contaminated water using pump-and-treat technology alone.

6.12. *Treatment/Disposal Solution-12.*

a. The adsorption data are plotted in the graph below.

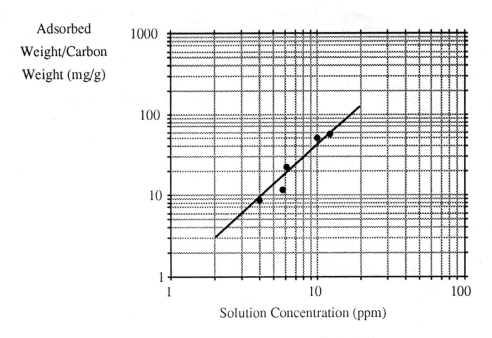

Taking the log of both sides of the Freundlich isotherm equation yields:

$$\ln\left(\frac{X}{M}\right) = \ln a + b \ln C$$

This is a linear equation of the form:

$$y = \text{constant} + m\,x$$

where $y = \ln(X/M)$, constant $= \ln a$, $m = b$, and $x = \ln C$. Least-squares linear regression analysis is used to determine the values of the slope, m, and the constant that best fit the experimental data. The resulting equation, after reconverting from the logarithmic form, is:

$$X/M = 1.02\,C^{1.56}$$

b. For 5 ppm, the adsorption capacity is:

$1.02\,(5)^{1.56} = 12.6$ mg benzene/g GAC

The mass flow of benzene is:

10 gpm (3.78 L/gal) (5 mg/L) = 189 mg/min

6.12. *Treatment/Disposal Solution-12 (continued).*

The daily GAC requirement is:

(189 mg benzene/min)/(12.6 mg benzene/g GAC)
= (15 g GAC/min) (1440 min/d)/(454 g/lb) = 47.6 lb GAC/d

The annual cost of GAC treatment, assuming the GAC is not regenerated, and fresh GAC is purchased, is then:

(47.6 lb/d)(365 d/yr)($7/lb) = $121,560/yr

6.13. *Treatment/Disposal Solution-13.*

For zero order kinetics,

$$dc/dt = -k \tag{1}$$

where, c = concentration of ethylacetate, g/m^3; t = residence time, sec; and k = rate constant, g/m^3-sec.

The rate constant, k, depends on the chemical to be treated, type of microorganisms on the biofilter, and biofilter operating conditions. For a packed bed column, residence time may be related to column height, and Equation 1 may be rewritten as,

$$u\, dc/dh = -k \tag{2}$$

where, u = superficial gas velocity, cm/sec; and h = packing height, cm. Equation 2 may be integrated to yield:

$$c = c_0 - k\,(h/u) \tag{3}$$

Here, c_0 is the concentration at $h = 0$.

Equation 3 may be rewritten as,

$$1 - c/c_0 = k\,[h/(uc_0)] \tag{4}$$

According to Equation 4, a plot of $1 - c/c_0$ versus $h/(uc_0)$ would yield a straight line through the origin with a slope of k. In the following table, y and x axes values for Equation 4 are obtained from the pilot plant studies:

6.13. *Treatment/Disposal Solution-13 (continued).*

Reactor #	y axis value $1-c/c_o$	x axis value $h/(uc_o)$
1	0.25	50
2	0.375	100
3	0.625	150
4	0.75	200
5	0.975	250

The value of k may be obtained by least squares regression techniques. The square of the error in the estimate e_i^2 is given by:

$$e_i^2 = (y_i - k\,x_i)^2$$

k is estimated by:

$$e_i^2/k = 0 = (y_i - k\,x_i)\,x_i$$

i.e., $k = y_i\,x_i/x_i^2 = 3.9 \times 10^{-3}$ g/m^3–s

The value of the rate constant, k, obtained from the pilot plant studies may be used to size the commercial reactor.

For the commercial reactor,

h = 1 m = 100 cm; c_o = 0.5 g/m^3; c/c_o = 0.9

Using the value of k previously obtained and Equation 4:

$1 - 0.9 = 3.9 \times 10^{-3}$ (100/u)
u = 7.84 cm/s (1 m/100 cm) (3600 s/h) = 282.2 m/h

The flow rate of the gas is 15,000 m^3/h. Therefore, the required biofilter area is:

(15,000 m^3/h)/(282.2 m/h) = 53 m^2.

6.14. *Treatment/Disposal Solution-14.*

a. From the particle size data presented in Table 1, the following particle size distribution histogram (Figure 2) can be generated for entering particles.

6.14. Treatment/Disposal Solution-14 (continued).

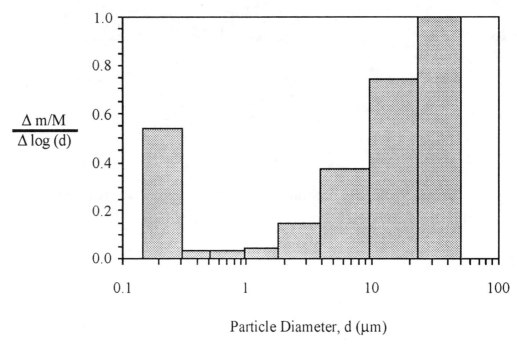

Figure 2.   Size distribution of particles entering the scrubber.

From the particle size data presented in Table 1, the following particle size distribution histogram (Figure 3) can be generated for leaving particles.

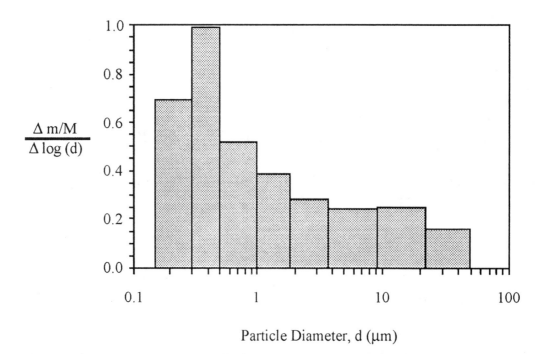

Figure 3.   Size distribution of particles leaving the scrubber.

6.14. *Treatment/Disposal   Solution-14   (continued).*

b. The size distribution of the particles entering the scrubber is bimodal.   The particles at the lower particle range are formed by nucleation and condensation.   The coarser particles are mainly non–volatile ash.

c. Using Equation 2 and data listed in Table 1, the mass mean diameters are:

Entering scrubber: d_m = 16.8 µm
Leaving scrubber: d_m = 4.2 µm

The mean diameter and size distribution have shifted towards the finer particle size.   The scrubber is size–selective, as a greater proportion of large particles than fine particles were removed.

d. Using Equation 3:

$$\eta_{overall} = (1 - 46.4/3402.6) \times 100 = 98.6 \ \%$$

e. The scrubber collection efficiency as a function of particle diameter is summarized in Figure 4 below.

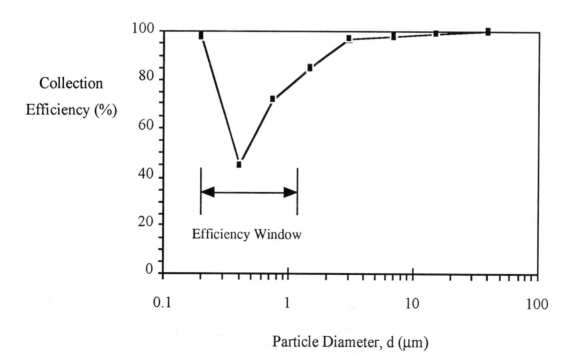

Figure 4.   Scrubber collection efficiency as a function of entering particle diameter.

6.14. *Treatment/Disposal Solution-14 (continued)*.

f. The scrubber does not remove particles uniformly at all particle diameters. Collection efficiency is very low for particles between 0.3 and 1 μm, but is high for particles outside this range. This efficiency window is characteristic of many gas cleaning devices.

Overall efficiency is based on the mass of particles removed, and is biased towards large particles. Fine particles are neglected because they constitute only a small proportion of the total mass. (Note that the term "fine" in this discussion is used to refer to particles between 0.3 and 1 μm.) However, fine particles are very important because of their impact on public health and because they cause visibility degradation in the atmosphere (haze). Emissions of large and very dense particles have a smaller effect because they settle to the ground relatively quickly, whereas fine particles persist in the atmosphere for longer periods.

Overall efficiency ($\eta_{overall}$) attempts to describe particle collection with a single number, but it does not characterize the entire size distribution. The efficiency curve is a more informative indicator of scrubber effectiveness.

Discussion

The chemical composition of aerosol emissions should also be considered. Industrial wastes often contain heavy metals, and in some cases inhaled metals produce more severe health effects than digested metals. Lead, nickel, cadmium, mercury and chromium are particularly toxic (Brunner, 1985). Particle composition varies with particle size, and size distributions should also be determined for particles containing hazardous substances.

6.15. *Treatment/Disposal Solution-15*.

a. The implications of Equation 1 depend on the value of D_f. Since the outer exponent affects all of the variables equally, it can be neglected for this analysis. If $D_f = 3$, then Equation 1 becomes:

$$d_m \propto \left(T^{0.5} \phi t \right)^{0.4}$$

6.15. *Treatment/Disposal Solution-15 (continued).*

In this case, primary particle size, d_0, has no effect on mean diameter, d_m. Both residence time, t, and particle loading, ϕ, are more important than temperature, T, as they have exponents of 1 (neglecting the outer exponent) whereas mean diameter, d_m, varies only with the square root of temperature, T.

If $D_f = 2$, then Equation 1 becomes:

$$d_m \propto T^{0.5} \, d_0^{-1.5} \, \phi \, t$$

Mean diameter, d_m, depends quite strongly on primary particle diameter, d_0, because of the relatively large negative exponent on d_0. A smaller primary particle size, d_0, results in a larger mean diameter, d_m.

b. The mean diameter, d_m, of the finer particles can be increased to reduce emissions. This shifts the size distribution of the fine particles to larger sizes, resulting in a greater proportion of particles which are larger than the efficiency window.

  i. Longer residence time, t, may be achieved by lengthening the flow path from the incinerator to the scrubber.
  ii. Temperature, T, has only a minor influence, but combustion conditions may possibly be adjusted to give higher flame temperatures.
  iii. Without further information, such as the factors which influence primary particle size, it is not possible to suggest a method to alter primary particle diameter, d_0. Also, the effect of primary particle diameter, d_0, is not significant if D_f is near 3
  iv. Particle loading, ϕ, is dependent on the composition of the waste being incinerated. Equation 1 applies to particles formed by nucleation and coagulation, and the amount of these particles which are produced depends on the amount of volatile condensable species in the sludge. Since municipal waste is composed of material from many sources it is unlikely that the composition can be controlled.

c. Other methods to modify particle size distribution include:

  i. Electrostatics, either by passing the aerosol through an electric field, or by adding an easily ionized material (such as sodium) to

6.15. *Treatment/Disposal Solution-15 (continued).*

the combustion process. Charged particles are subject to additional attractive forces that enhance agglomeration.

ii. Acoustic agglomeration, using high intensity sound waves.
iii. Add coarse particles which can scavenge fine particles.
iv. Make the flow more turbulent.

Discussion

i. Suggested ways of recycling incinerator ash include the following. Note that some wastes may be hazardous and therefore non-recyclable.

    i. Structural fill – for construction, land reclamation, embankments, etc.
    ii. Mineral wool manufacture – utilizing the glass fraction of combustion residue.
    iii. Extender – utilized for the production of pavement, concrete, brick, etc.

ii. Aerosol formation can be prevented by selecting a feed material which will form only gaseous products, and optimizing combustion conditions so that aerosol by-products cannot be formed. For example, coal can be pretreated to remove mineral matter (benefaction). Other coal processing techniques produce fuels which do not contain minerals (e.g., liquefaction). In waste incineration, waste streams can be characterized to prevent inorganics, such as heavy metals, from being incinerated.

7. Chemical Plant/Domestic Applications

7.1. *Chemical Plant/Domestic Applications Solution-1.*

a. This is a saturated solution with no vapor formed so the total pressure is the sum of the pressures exerted by the components in the liquid phase, i.e.,

$$P = x_1 P_1 + x_2 P_2 = 0.05 \, (4150) + 0.95 \, (16.1) = 222.8 \text{ kPa}$$

The vapor composition for the first bubble formed will be:

$$y_1 = x_1 P_1/P = 0.05 \, (4150)/222.8 = 0.931$$

b. There are several forms of the flash equations. The one most frequently used for numerical solution is:

$$\sum_{i=1}^{n} (z_i(1 - K_i)/(1+v(K_i - 1))) = 0$$

Here z_i is the gmol fraction of component i in the feed, v is the fraction of feed vaporized, K_i is y_i/x_i or P_i/P for an ideal gas; and n is the total number of components.

For a binary mixture, n = 2 so:

$$z_1 (1 - K_1)/(1 + v (K_1 - 1)) + z_2 (1 - K_2)/(1 + v (K_2 - 1)) = 0$$

Here $z_1 = 0.05$, $z_2 = 0.95$, $K_1 = 4150/P$, $K_2 = 16.1/P$, and $v = 0.10$

This is a quatratic equation in P, but a trial and error solution is easier. This solution gives P = 30.2 kPa. From which:

$$x_1 = z_1/(1 + v(K_1 - 1)) = 0.05/(1 + 0.1(4150/30.2 - 1)) = 0.003415$$

and $x_2 = 1 - x_1 = 0.9966$. Then $y_1 = K_1 x_1 = 0.4693$ and $y_2 = 0.5307$. The removal of ethane is the amount in the vapor divided by the amount in the feed or $v \, y_1/Z_1 = 0.1 \, (0.4693/0.05) = 93.8\%$

7.2. *Chemical Plant/Domestic Applications Solution-2.*

a. First determine the total number of gmol of air in the room. Keep in mind that for an ideal gas 1 gmol occupies 22.4 L (or 0.0224 m³) at STP (273 K, 1 atm).

7.2. *Chemical Plant/Domestic Applications Solution-2 (continued).*

The room temperature is at 22°C = 295 K

Total air gmol in room = (1100 m³) (1000 gmol/22.4 STP m³) (273 K/295 K)
    = 45,445 gmol air

The mole fraction of HC = 1.5 gmol HC/(45,445 gmol air + 1.5 gmol HC)
    = 0.000033 gmol of HC/gmol of mixture = 33 ppm
    = 33,000 ppb >> 850 ppb. Yes, there is a health risk!

b. Have the potential rupture area vented directly into a hood or a duct to capture any leakage in case of rupture.

c. Input substitution: possibly replace the material in the reactor with material that would have a lower vapor pressure.

7.3. *Chemical Plant/Domestic Applications Solution-3.*

a. The vapor volume in m³/kgmol is:

$$V = RT/P = 8.314 \ (293)/101.3 = 24.05 \ m^3/kgmol$$

where R = 8.314 m³-kPa/kgmol-K

The gasoline in the vapor is:

10 gal (m³/264.1 gal)(kgmol/24.05 m³) (0.40) = 0.000630 kgmol

The liquid volume of the gasoline is:

0.000630 kgmol (70 kg/kgmol) (m³/1000 kg) (1/0.62) (264.1 gal/m³)
    = 0.0188 gal

b. The gasoline loss per car per year is:

0.0188 gal/fill (52 fills/yr) = 0.98 gal/car/yr

For 50 million cars this becomes 48,900,000 gal/yr.

c. The value of the gasoline lost is:

(48,900,000 gal/yr) ($1.20/gal) = $58,700,000/yr

7.4. *Chemical Plant/Domestic Applications Solution-4.*

a. The rate expression for D is:

$$\frac{d[D]}{dt} = k_1 [A] - k_2 [A] [D] + k_3 [U]$$

The rate expression of U is:

$$\frac{d[U]}{dt} = k_2 [A] [D] - k_3 [U]$$

b. This solution was carried out using a differential equations solver on the Macintosh (STELLA II), that is able to generate tabular and graphical output of the system of differential equations presented above. With the constants and initial conditions given in the problem statement, the following results were obtained for $[A]_0 = 1$ and $[D]_0 = [U]_0 = 0$.

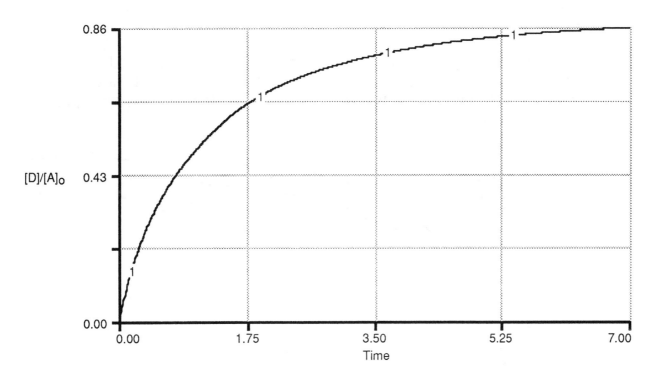

Figure 1. $[D]/[A]_0$ versus time for reaction scheme given in problem statement, with $[A]_0 = 1$ and $[D]_0 = [U] = 0$.

7.4. Chemical Plant/Domestic Applications Solution-4 (continued).

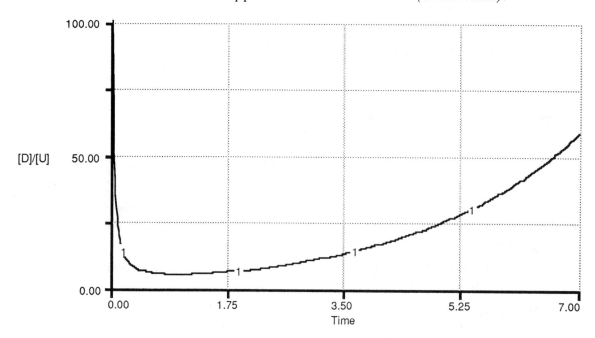

Figure 2. [D]/[U] versus time for reaction scheme given in problem
statement, with $[A]_0 = 1$ and $[D]_0 = [U] = 0$.

c. This solution was carried out using STELLA II, in the simulation
   mode, to generate output as a function of variable initial values of
   $[A]_0$. The results are summarized for $[A]_0 = 1$, 0.5 and 0.1, with $[D]_0 = [U]_0 = 0$.

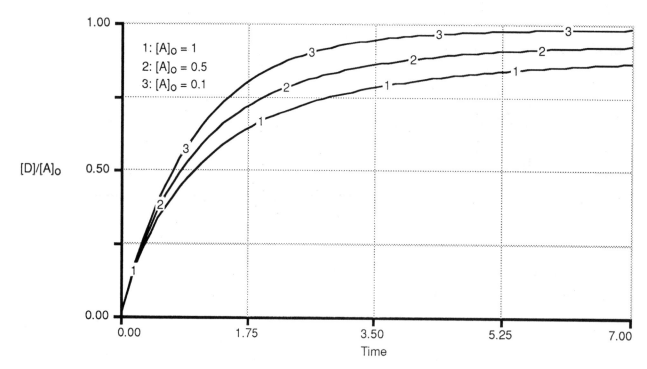

Figure 3. $[D]/[A]_0$ versus time for reaction scheme given in problem
statement, with $[A]_0 = 1$, 0.5 and 0.1, with $[D]_0 = [U]_0 = 0$.

7.4. *Chemical Plant/Domestic Applications Solution-4 (continued).*

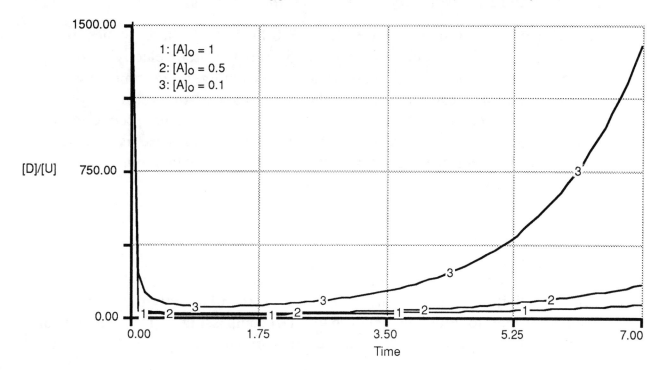

Figure 4.   [D]/[U] versus time for reaction scheme given in problem statement, with $[A]_0 = 1$, 0.5 and 0.1, with $[D]_0 = [U]_0 = 0$.

d. For 85% conversion of A to D, the selectivity, the ratio of [D]/[U], is found from Figure 4 for each value of $[A]_0$ as follows:

$[A]_0$	1.0	0.5	0.1
[D]/[U]	32.5	21.8	59.8

As the reactant is diluted above 50%, the reaction selectivity favors the desired product. This suggests that large dilutions by an inert species can be used as a source reduction procedure by reducing the formation of undesired by-products using this particular reaction scheme, and the given rate constants and initial conditions.

e. At 85% conversion, the selectivity is given in Part d and the batch process times, excluding down-time, are determined from Figure 3 to be:

$[A]_0$	1.0	0.5	0.1
t	5.6	3.2	2.1

7.4. *Chemical Plant/Domestic Applications Solution-4 (continued).*

Comparing the most diluted case ($[A_0]$ = 0.1) with the most concentrated case ($[A_0]$ = 1.0) yields 1.8 (59.8/32.5) times better selectivity using 37.5% (2.1/5.6) shorter process times. Assuming three times as many batches/yr can be run for the diluted case, the reactor volume will need to be 3.33 times greater to achieve the same production rate. Also, the waste stream volume for the diluted case is larger, which will affect the treatment load if other pollutants must be removed from it.

7.5. *Chemical Plant/Domestic Applications Solution-5.*

This is an open-ended question that is designed to promote awareness and creativity in the student. The use of familiar activities can sensitize one to thinking small but all-encompassing. The transfer of the attitudes developed here are easily transferred to more specific academic and industrial situations.

The types of pollution that the student should consider at each stage include:

Air - chemicals and particulates, greenhouse effect, depletion of the ozone layer, acid rain
Water - organic and inorganic chemicals, load on the water treatment plant, ground water, surface water, oceans, acid rain
Land - organic and inorganic chemicals, landfill requirements

and two which are rarely considered, Thermal Pollution and Noise Pollution.

7.6. *Chemical Plant/Domestic Applications Solution-6.*

a. The annual cost for conventional AA batteries under the conditions stated will be:

4 batteries/mo (12 mo/yr) ($0.89/battery) = $42.72/yr

The annual cost for rechargeable Ni-Cad batteries for the first five years is:

[(8 batteries) ($2.75/battery) + $14/charger)]/5 yr
= $36/5 yr = $7.20/yr

7.6. *Chemical Plant/Domestic Applications Solution-6 (continued).*

b. Payback of the original capital cost of $36 would occur in:

$36/[(4 batteries/mo) ($0.89/battery)]   =   10.1 months

After 10 months, the solar-rechargeable system is free, at least until replacement is required. If the claim of 1,000 charges is accurate, then 5 yr is a conservative estimate. Obviously, if they last more than 5 yr, then the economics improve. Assuming the batteries last for 1,000 charges and the charger lasts indefinitely (not unreasonable) then the cost for four "new" batteries becomes less than $0.04, or a penny each. The cost of conventional batteries will not change (assuming no fluctuation in price).

As this problem illustrates, it makes good economic and environmental sense to switch to solar rechargeable batteries. There will be some inconvenience due to the necessity of charging the spent batteries, but not unlike having to run to the store (more wasted energy) to buy new conventional batteries. Of course, if the battery charger is left out in the rain, you may need a new one.

A way to enhance the economic picture is to recognize that the acts of not polluting and of not using resources is worth money. However, it is difficult to place a dollar value on this.

It is important to remember that eventually even the rechargeable batteries will need to be replaced and the spent batteries should not be disposed of in the household trash.

7.7. *Chemical Plant/Domestic Applications Solution-7.*

a. 1. adsorbtion control unit;
   2. freeboard
   3. water jacket
   4. heating elements
   5. cleanout door.

b. 1. roof vent
   2. surface diffusion and convection
   3. "drag-out" losses
   4. waste solvent
   5. leaks

7.7. *Chemical Plant/Domestic Applications Solution-7 (continued).*

   c. 1. b
       2. d
       3. c
       4. d
       5. b

   d. 1. True
       2. True
       3. False
       4. False
       5. True

7.8. *Chemical Plant/Domestic Applications Solution-8.*

   a. The mole fraction of gasoline vapor equivalent to 35 mg/L is:

(35mg/L) (1g/1000 mg) (1gmol/70 g of gas) (22.4 L/gmol total) (535 °R/492 °R)
= 0.0122 gmol gas/gmol total = vapor mole fraction

The partial pressure is y_P = 0.0122 (1 atm) = 0.0122 atm.

   b. The total pressure required to reduce the vapor concentration of gasoline to 35 mg/L at 75°F is:

$P = x P_v/y$ = (1) (6.5)/0.0122 = 533 psia = 36.2 atm

This compression pressure is probably too high for practical application and refrigeration should be used to lower the temperature or some type of absorption or adsorption process should be used.

   c. The recovery can be calculated by a material balance around the condenser. A total balance can be written as:

$$w_1 = w_2 + w_3$$

while a gasoline balance takes the form:

$$w_1 y_1 = w_2 y_2 + w_3 y_3$$

7.8. *Chemical Plant/Domestic Applications Solution-8 (continued).*

Here $w_1 = 1$ gmol of original vapor; $y_1 = x P_v/P = (1) (6.7)/14.7 = 0.442 =$ mole fraction of gasoline in the original vapor; $w_2 =$ gmol of gasoline condensed; $y_2 =$ mole fraction of condensed gasoline $= 1.0$; $w_3 =$ gmol in the residual vapor stream; and $y_3 = 0.0122 =$ the mole fraction of gasoline in the residual vapor stream.

Solution of these equations for w_2 gives $w_2 = 0.435$, and $w_3 = 0.565$. The recovery is then $0.435/0.442 = 0.984$.

7.9. *Chemical Plant/Domestic Applications Solution-9.*

a. For a total PCE demand of 4,000 lb/yr, the fresh PCE demand is found from a mass balance around the degreaser provided in the solution to Problem 1.4. Basic Concepts - 4. The solution of Problem 1.4 shows that for 1 lb of fresh PCE feed, the recovered PCE stream is 0.2 lb, and the PCE emission stream is 0.94 lb. For a total PCE demand (fresh PCE + recovered PCE) of 4000 lb/yr, the fresh PCE requirement is calculated as:

Input = 4,000 lb PCE/yr = Fresh + Recycled PCE
   = Fresh PCE + 0.2 Fresh PCE = 1.2 (Fresh PCE)
Therefore,

Fresh PCE = (4,000 lb PCE/yr)/(1.2) = 3333 lb Fresh PCE/yr, and:

PCE Emissions = (0.94 lb/lb Fresh PCE) (3333 lb Fresh PCE/yr)
   = 3133 lb PCE Emissions/yr

b. With 90% solvent recovery efficiency, the mass balances around the solvent recovery unit and entire system allow the determination of a new value for the amount of recycle PCE, X.

The Solvent Recovery mass balance yields:

Spent PCE = Recovered PCE + Rejected PCE
0.22 (1 + X) = 0.90 (0.22 (1 + X)) + Rejected PCE
Rejected PCE = 0.022 (1 + X) = 0.022 + 0.022 X

The total system mass balance yields:

7.9. *Chemical Plant/Domestic Applications Solution-9 (continued).*

Fresh PCE = PCE Emissions + Rejected PCE
1 lb PCE = (0.78 + 0.78 X) + (0.022 + 0.022 X) = 0.802 + 0.802 X
X = 0.198/0.802 = 0.247 = 0.25 lb PCE

PCE Emissions = 0.78 + 0.78 (0.25) = 0.975 lb/lb Fresh PCE

With 90% solvent recovery efficiency, for 1 lb of fresh PCE feed, the recoverd PCE stream is 0.90 (0.22 (1 + 0.25)) = 0.2475 lb, and the PCE emission stream is 0.975 lb. For a total PCE demand (fresh PCE + recovered PCE) of 4000 lb/yr, the fresh PCE requirement is therefore:

4,000 lb PCE/yr = Fresh + Recycled PCE = Fresh PCE + 0.25 Fresh PCE
= 1.25 (Fresh PCE)

Threrefore,

Fresh PCE = (4000 lb/yr)/1.25 = 3200 lb Fresh PCE/yr, and:

PCE Emissions = (0.975 lb/lb Fresh PCE) (3200 lb Fresh PCE/yr)
= 3120 lb PCE Emissions/yr

This amounts to a savings = 3133 - 3120 = 13 lb PCE/yr

c. The emission rate for a 25% efficient vapor recovery unit is determined from a new mass balance for the entire system:

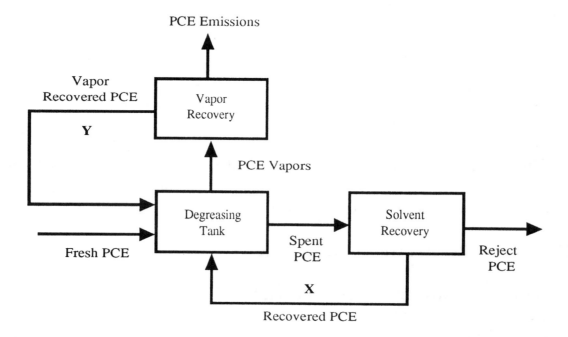

7.9. *Chemical Plant/Domestic Applications Solution-9 (continued).*

The mass balance around the degreasing tank using a 1 lb fresh PCE basis is as follows:

Input = Fresh + Recovered PCE + Vapor Recovered PCE
= 1 + X + Y lb PCE

Output = PCE Vapors + Spent PCE = 0.78 (1 + X + Y) + Spent PCE

Input = Output; 1 + X + Y lb PCE = 0.78 (1 + X + Y) + Spent PCE

Spent PCE = (1 - 0.78) (1 + X + Y) = 0.22 (1 + X + Y)
= 0.22 + 0.22 X + 0.22 Y lb PCE

The mass balance around the vapor recovery unit is as follows:

Input = PCE Vapors = 0.78 (1 + X + Y) lb PCE

Output = Vapor Recovered PCE + PCE Emissions
Output = 0.25 [0.78 (1 + X + Y)] + PCE Emissions

Input = Output
0.78 (1 + X + Y) = 0.25 [0.78 (1 + X + Y)] + PCE Emissions
PCE Emissions = (1 - 0.25) [0.78 (1 + X + Y)] = 0.585 (1 + X + Y)
= 0.585 + 0.585 X + 0.585 Y lb PCE

The mass balance around the solvent recovery unit is as follows:

Input = Spent PCE = 0.22 + 0.22 X + 0.22 Y lb PCE

Output = Recovered PCE + Rejected PCE
Recovered PCE = 75% of Spent PCE
Output = 0.75 (0.22 + 0.22 X + 0.22 Y lb PCE) + Rejected PCE

Input = Output
0.22 + 0.22 X + 0.22 Y = 0.75 (0.22 + 0.22 X + 0.22 Y lb PCE)
+ Rejected PCE
Rejected PCE = (1 - 0.75) (0.22 + 0.22 X + 0.22 Y)
= 0.055 + 0.055 X + 0.055 Y lb

The mass balance around the entire system is as follows:

7.9. *Chemical Plant/Domestic Applications Solution-9 (continued).*

Input = 1 lb PCE

Output = PCE Emissions + Spent PCE
   = (0.585 + 0.585 X + 0.585 Y) + (0.055 + 0.055 X + 0.055 Y)

Input = Output
1 lb PCE = (0.585 + 0.585 X + 0.585 Y) + (0.055 + 0.055 X + 0.055 Y)
1 lb PCE = 0.64 + 0.64 X + 0.64 Y
0.36 = 0.64 X + 0.64 Y

The Recovered PCE = X = 0.75 (0.22 + 0.22 X + 0.22 Y)
   = 0.165 + 0.165 X + 0.165 Y lb PCE
   0.835 X = 0.165 + 0.165 Y; X = 0.198 + 0.198 Y

Substituting this result into the equation from the entire system mass balance yields the following:

0.36 = 0.64 (0.198 + 0.198 Y) + 0.64 Y = 0.127 + 0.127 Y + 0.64 Y
Y = 0.233/0.767 = 0.304 lb Vapor Recovered PCE/lb Fresh PCE

Then X = 0.198 + 0.198 Y = 0.198 + 0.198 (0.304)
   = 0.258 lb Solvent Recovered PCE/lb Fresh PCE

The amount of Fresh PCE required is:

Fresh PCE = (4000 lb/yr)/(1 + 0.258 + 0.304)
   = 4000/1.562 = 2561 lb/yr

The lb PCE emitted/lb fresh PCE utilized from the mass balance calculations is:

PCE Emissions = 0.585 + 0.585 X + 0.585 Y
   = 0.585 + 0.585 (0.258) + 0.585 (0.304)
   = 0.914 lb PCE Emitted/lb fresh PCE

Total PCE Emissions are then:

PCE Emissions = 0.914 (2561) = 2341 lb PCE Emissions/yr

7.9. *Chemical Plant/Domestic Applications Solution-9 (continued).*

The savings for this option = 3,133 - 2,341 = 792 lb PCE/yr

d. Even with a highly inefficient vapor recovery system, the PCE emission reduction is greater than 60 times more effective than the higher efficiency solvent recovery system. With higher efficiency vapor recovery, the savings and emission reductions would be even more dramatic. Evaluation and implementation of vapor recovery would be the most productive route to pursue with management to provide a significant, cost effective reduction in PCE emission rates from this degreaser.

7.10. *Chemical Plant/Domestic Applications Solution-10.*

a. The extent of vapor losses are directly proportional to the mass flux of solvent vapor (S) into air (A):

$$N_{SA} = \frac{D_{SA} P}{RT (h_2 - h_1)} \ln\frac{p_2}{p_1} \tag{1}$$

D_{SA}, P/RT, $h_2 - h_1$ and p_2 are constant for the system. Equation 1 may be rewritten as:

$$N_{SA} = k \ln(p_2/p_1)$$

The flux is reduced by 50%, therefore:

$$0.5 \ N_{SA} = k \ln(p_2/p_1^*)$$

where p_1^* is the new pressure at the interface.

Thus,

$$\ln(p_2/p_1) = 2 \ln(p_2/p_1^*), \text{ or } p_1^* = (p_2 p_1)^{0.5}$$

From the Antoine equation given in the problem statement:

$$\ln (p_1) = A - \frac{B}{(T_1 + C)} \tag{2}$$

With input values from Table 1:

7.10. *Chemical Plant/Domestic Applications Solution-10 (continued).*

$$\ln(p_1) = 16.2 - \frac{3028}{\left(\left(273.2 + \frac{87.2}{2}\right) - 43.2\right)} = 5.13$$

$p_1 = 169.5$ mm; and $p_1{}^* = 65.1$ mm

The corresponding temperature is:

$$\ln(65.1) = 16.2 - \frac{3028}{(T - 43.2)}$$

$T = 22°C$

b. The mass flux of solvent vapor (S) into air (A) is described by the equation presented in Part a. As per Part a, D_{SA}, P/RT, $h_2 - h_1$ and p_2 are constant for the system.

The flux is reduced by 50%. If $p_1{}^{min}$ is the new pressure at the interface for the TCA system, then:

$$p_1{}^{min} = (p_2 \, p_1{}^{TCA})^{0.5}$$

From the Antoine equation:

$$\ln(p_1{}^{TCA}) = A - \frac{B}{(T_1 + C)}$$

From Table 1 for the solvent TCA, the interface pressure at 50% below the boiling point of TCA is estimated as:

$$\ln(p_1{}^{TCA}) = 16.0 - \frac{3110}{\left(\left(273.2 + \frac{113.7}{2}\right) - 56.2\right)} = 4.63$$

$p_1{}^{TCA} = 102.8$ mm, and therefore, $p_1{}^{min} = 50.7$ mm

The required temperature from the Antoine equation is:

$$\ln(50.7) = 16.0 - \frac{3110}{(T - 56.2)}$$

$T = 40.6°C$

On the basis of vapor pressure alone, TCA would be preferred over TCE. However, energy costs and long term atmospheric effects should be included in the considerations.

7.10. *Chemical Plant/Domestic Applications Solution-10 (continued).*

c. The boiling point of TCE = 87.2°C and the lowest allowable temperature at the interface is 25% of the boiling point in °C

$$bp^{lowest} = (0.25)(87.2) = 21.8°C$$

The corresponding pressure p_1^{**} in mm at the interface is given by:

$$\ln(p_1^{**}) = A - \frac{B}{(T_1 + C)}$$

From Table 1

$$\ln(p_1^{**}) = 16.2 - \frac{3028}{((273.2 + 21.8) - 43.2)} = 4.175$$

$$p_1^{**} = 65 \text{ mm.}$$

The maximum possible reduction in vapor loss is estimated from Part a as:

$$N_{SA} = k \ln(p_2/p_1), \text{ where } p_1 = 169.5 \text{ mm and } p_2 = 25 \text{ mm}$$

The flux is reduced by a factor z (to be determined), therefore:

$$z N_{SA} = k \ln(p_2/p_1^{**})$$

where $p_1^{**} = 65$ mm (the new pressure at the interface, from above).

Equating the two flux conditions yields:

$$z \ln(p_2/p_1) = \ln(p_2/p_1^{**})$$

or

$$z = [\ln(p_2/p_1)]/[\ln(p_2/p_1^{**})] = [\ln(25/169.5)]/[\ln(25/65)] = 0.50$$

The interpretation is that the vapor losses in this system cannot be reduced by more than 50%.

d. In an open tank, air currents cause turbulence in the vapor zone. The mixing of air with solvent vapors reduces the condensation efficiency, resulting in increased solvent losses. The main function of the freeboard is to minimize these losses.

7.11. *Chemical Plant/Domestic Applications Solution-11.*

a. Membrane 1, a hydrophyllic polymer, displays better rejection of contaminant AOH than Membrane 2. From Figure 2, the optimum operating pressure P is 14×10^3 torr.

The AOH flux at the operating pressure from Figure 2 is approximately 0.0012 L/m²–h.

The permeate flowrate, extrapolated from the H_2O line in Figure 2, is 15 L/m²–h.

The concentration of AOH in the permeate
= (0.0012 L/m²–h)/(15 L/m²–h)
= 0.00008 = 0.008 %
= 80 mg/L since the specific gravity of AOH is 1.00.

b. The concentration ratio for AOH = final concentration/initial concentration
= (20 g/L)/(0.5 g/L) = 40

The final volume flowrate of the AOH "product"= initial volume flowrate/concentration ratio
= (1,000 L/d)/40 = 25 L/d

The number of separation times is determined from the permeate flowrate/permeate flowrate per RO unit, or:

Required separation units
= (1000 L/d – 25 L/d)/[(15 L/hr-unit) (24 h/d)] = 3 units

c. From Figure 2, the optimal membrane appears to be hydrophyllic in nature; thus a high degree of rejection of the ionized species is expected. Since the organic contaminant ionizes as follows:

$$AOH \rightleftarrows [AO^-] + [H^+]$$

$$K_a = [AO^-][H^+]/AOH = 8 \times 10^{-11}$$

an operating pH>>pK_a is necessary for complete ionization. A pH range of 10 to 12 would be preferred.

7.12. *Chemical Plant/Domestic Applications Solution-12.*

The annual depreciation is:

Depreciation = $7,000,000/6 yr = $1,170,000/yr

Yearly income = Sales - Manufacturing Costs - Depreciation
= $4,700,000 - $1,400,000 - $1,170,000 = $2,140,000/yr

The rate of return on the investment = yearly income/total investment
= $2,140,000/$7,000,000 = 0.31 or 31% $/yr-$

The payback period = total investment/yearly income
= $7,000,000/$2,140,000 = 3.3 yr

This is an excellent rate of return on investment, and the company should seriously consider investing in this enterprise.

7.13. *Chemical Plant/Domestic Applications Solution-13.*

The total number of tires processed = (260 d/yr) (24 h/d) (440 tire/h)
= 2,745,600 tires/yr

Crude oil production = (2,745,600 tires/yr) (1.5 gal/tire)
= 4,118,400 gal/yr

Carbon black production = (2,745,600 tires/yr) (5 lb/tire)
= 13,728,000 lb/yr = 6,240 ton/yr

Total Investment = $6,000,000 + $500,000 + $500,000 = $7,000,000

Depreciation = (Fixed investment - salvage value)/economic life
= ($6,000,000 - 0)/6 = $1,000,000/yr

For a zero net profit, P = 0, then the gross profit rate, R, is:

$R = e\ IF + (R - d\ IF)\ t$
$R\ (1-t) = e\ IF\ (1 - t)$
$R = e\ IF = IF/N = \$1,000,000/yr = S - C = S - \$1,000,000$
$S = \$2,000,000/yr$

The annual sales has to be $2,000,000/yr.

7.13. *Chemical Plant/Domestic Applications Solution-13 (continued).*

The cost of crude oil = ($2,000,000/yr)/(4.118 Mgal/yr)
  = $0.486/gal

If carbon black is sold at $350/ton, the revenues are:

($350/ton)(6240 ton/yr) = $2,184,000

This is more than the annual sales revenue (without profit). Thus, the oil is produced at a zero cost under this scenario.

8. Case Studies

8.1. *Case Studies Solution - 1.*

a. Assume that no neutralization takes place. The only effect is the dilution of the acid by the approximately 100 gal (or $\approx$ 400 L) of water in the tank. Remember that:

$$pH = -\log [H^+] \text{ and } [H^+] = 10\text{-}pH$$

If the tank has a pH = 2.0, then $[H^+] = 10^{-2}$. The total number of gmol of H^+ in the 400 L volume would then be:

$$(400 \text{ L}) (0.010 \text{ gmol } H^+/L) = 4.0 \text{ gmol } H^+$$

That is, 4.0 gmol of H^+ put down the drain would bring the pH in the tank to 2.0. Since concentrated HCl contains 12 gmol H^+/L, and approximately 100% of it is ionized in water, 0.333 L would contain the requisite 4.0 gmol H^+.

b. Since H_2SO_4 has 2 acidic hydrogens, and approximately 100% of it is ionized in water, its $[H^+]$ is double its molarity. Thus, concentrated (18 molar) H_2SO_4 would contain 36 gmol H^+/L. One-ninth of a L (or 111 mL) of 18 molar H_2SO_4 would contain the 4.0 gmol of H^+ needed to bring the tank to a pH of 2.0.

c. More than one correct answer to this question is possible. The college found the following actions to be effective:

1. <u>Source Reduction</u>. All experiments using acids were reviewed and revised to use the minimum amount of acid necessary to successfully carry out the desired learning experience.
2. <u>Dilution</u>. Two sinks in each lab were fitted with "drain extenders" (that is a pipe that fits into the drain and causes the sink to fill with water before draining). Water was allowed to flow into these sinks continuously during laboratory sessions. Students were instructed to pour all waste acid into these sinks. This substantially diluted the acid before it entered the tank. For those experiments where these two actions were inadequate, additional water was added to the drains during those laboratory periods by running water continuously in other sinks.

NOTE: The college found that neutralization of the acid before pouring it down the drain required the use of strong bases in order for the

8.1. *Case Studies Solution - 1 (continued).*

> process to be fast enough to be completed during the laboratory sessions. The college decided not to do these neutralizations because of the safety problems associated with them.

8.2. *Case Studies Solution - 2.*

> This is a mass balance problem. The mass of chromium disposed will be the amount that is present in the dewatered sludge. First estimate the amount of dewatered sludge from the given data.

The amount of filtrate/d
   = filter area (volume of filtrate/hr-ft^2) (h of operation/d)
   = 150 ft^2 (10 gal/hr-ft^2) (16 h/d)   = 24,000 gal/d

Develop mass balance equations relating the raw sludge and the dewatered sludge as follows:

Mass Balance Equation 1:

Mass of raw sludge = mass of dewatered sludge + mass of filtrate

Mass Balance Equation 2:

Mass of solids in the raw sludge = mass of solids in the dewatered sludge

With these equations, the following input data are available:

mass of raw sludge not known
mass of dewatered sludge not known
mass of filtrate = 24,000 gal/d (8.34 lb/gal) = 200,000 lb/d
mass of solids in the sludge = 0.05 (mass of raw sludge)
mass of solids in the dewatered sludge = 0.47 (mass of dewatered sludge)

Dividing Equation 1 by Equation 2 yields:

$$\frac{\text{mass of raw sludge}}{\text{mass of solids in raw sludge}} = \frac{\text{mass of dewatered sludge + filtrate}}{\text{mass of solids in dewatered sludge}}$$

8.2. *Case Studies Solution - 2 (continued).*

or,

0.05 = 0.47(mass of dewatered sludge, lb/d)/(mass of dewatered sludge per d + 200,160 lb/d)

dewatered sludge = 0.05 (200,000 lb/d)/(0.42) = 23,830 lb/d

The amount of total chromium in sludge is determined from the mass concentration of chromium in the sludge as:

mass of dewatered sludge (mass concentration of chromium in sludge) = 23,830 lb/d (120 lb/1,000,000 lb) = 2.86 lb/d

The chromium disposed/yr = 2.86 lb/d (7d/wk) (50 wk/yr) =1000 lb/yr

8.3. *Case Studies Solution - 3.*

a. The task is to first develop a mass balance around the distillation column.

$$\text{Input (F)} = \text{Output (L + G + H)}$$

where L=lights, G=oil/gas, H=heavy distillate

F = 1000 bbl/hr (42 gal/bbl) [(8.33)(0.958) lb/gal] = 335,166 lb/hr
L = 2240 ft³/min (0.185 lb/ft³)(60 min/hr) = 24,864 lb/hr
G = 457.1 bbl/hr (42 gal/bbl) [(8.34)(0.81) lb/gal] = 129,536 lb/hr
H = F - (L + G) = 180,766 lb/hr
H = 180,766 lb/hr/[((8.34) (1.15) lb/gal) (42 gal/bbl)]
    = 449.3 bbl/hr

b. The energy required in BTU/bbl for heating heavy distillates from 500°F to 950°F is calculated as follows:

$$q = \int_{T_1}^{T_2} C_p \, dT = \int_{T_1}^{T_2} (45.2 + 9 \times 10^{-2} \, T) \, s \, dT$$

8.3. *Case Studies Solution - 3 (continued).*

$$= (45.2 (T_2 - T_1) + 9 \times 10^{-2} (T_2^2 - T_1^2)/2) \, s$$
$$= (45.2 (450) + 9 \times 10^{-2} (1410^2 - 960^2)/2) \, s$$
$$= (20,340 + 89,464 - 41,472) \, s = 68,330 \, s \text{ Btu/bbl}$$

With $s = 1.15$, the energy requirement is:

$$q = 68,330 \, (1.15) = 78,580 \text{ Btu/bbl}$$

The total energy required to heat the heavy distillate is:

$$Q = (q \text{ Btu/bbl}) \, (\text{bbl/hr}) = 78,580 \text{ Btu/bbl} \, (449.3 \text{ bbl/hr})$$
$$= 35.3 \times 10^6 \text{ Btu/hr}$$

c. The volume of natural gas required/d is estimated based on the total energy requirement, the energy content of the natural gas, and the efficiency of utilization (80%) as:

$$(35.3 \times 10^6 \text{ Btu/hr})/[(940 \text{ Btu/ft}^3) \, (0.8)] = 46,940 \text{ ft}^3/\text{hr}$$

The equivalent natural gas (energy) cost is:

$$(46,940 \text{ ft}^3/\text{hr}) \, (\$1.50/1000 \text{ ft}^3) = \$70.41/\text{hr} = \$1690/\text{d}$$

d. For the first order hydrodesulfurization reaction

$$dC/dt = -kC, \text{ or upon integration, } \ln(C_i/C_t) = kt$$

where C_i = initial concentration of S, C_t = concentration of S at time t, t of interest in the problem is the residence time, V/v, V = reactor volume, and v = volumetric flowrate of the heavy distillate.

Using this formulation, the required reactor volume is found to be:

$$\ln(C_i/C_t) = kt = (0.4/\text{hr}) \, [(V \text{ gal})/(18,870 \text{ gal/hr})]$$
$$V = [18,870 \text{ gal/hr} \, (\ln(10/0.5))]/(0.4/\text{hr}) = 141,320 \text{ gal}$$
$$V \text{ corrected for } 20\% \text{ bed expansion} = 170,000 \text{ gal}$$

e. The first task is to determine the tons of SO_2 that would be generated/d without the hydrodesulfurizer. Since the amount of S is reduced from 10% to 0.5% from Part a,

8.3. *Case Studies Solution - 3 (continued).*

> S tons/d = (180,754 lb/hr) (24 hr/d) (0.1 - 0.005)/(2000 lb/ton)
> = 206 tons
> SO_2 = 2 (206 tons/d) = 412 tons/d, since molar ratio of S to O_2 is 1:1.

> The cost of operating pollution control equipment/yr (365 d) is estimated as:

> (412 ton/d) (365 d/yr) ($5.00/ton) = $752,500/yr

> The annual cost of the hydrodesulfurizer unit, including operation and energy costs, is:

> $0.235/gal (170,000 gal) + 0.3 [$0.235/gal (170,000 gal)] + ($1690/d) (365 d/yr)
> = $670,610/yr

> The cost of the two processes appears to be comparable. The hydrodesulfurizer unit has a slight cost advantage of approximately 11%. It would be the preferred alternative since the early removal of sulfur is preferred to avoid having to clean-up sidestreams that also contain sulfur.

8.4. *Case Studies Solution - 4.*

> First, identify pollution prevention measures especially for the rinse water. Secondly, a cost analysis of the recommended pollution prevention methods needs to be conducted.

> Pollution prevention approach for the rinse water: Two options that can be considered for the rinse water are: (1) to reduce the flow, and (2) convert to a batch rinse.

> Based on the drag-out information provided, the total amount of drag-out
> = board area (drag-out rate) = 2.5 ft^2 (15 mL/ft^2) = 37.5 mL
> =0.375 L (0.2642 gal/L) = 0.0099 gal

8.4. *Case Studies Solution - 4 (continued).*

With the general industry standard of 1/1000 concentration in the rinse tanks, the volume of water required to rinse this piece of 2.5 ft^2 circuit board is:

37.5mL (1000/1) = 37,500 mL = 9.9 gal

Only 10 gal of rinse water are needed to meet the industry standard.

Thus, a 30-gal rinse water tank will provide adequate rinsing without needing to operate the tank as a flow-through unit. Thus a reduction of 16 gpm of water per rinse tank can be realized. However, in some cases the tank has to operated on a flow-through basis because of the chemistry involved. Even then the amount of water required (flow rate) can be substantially reduced.

Based on flow rate of rinse water, the concentration ratio can be estimated for rinse the tanks as follows:

$$Q = D (C_b/C_r)/t$$

where Q = rinse tank flow rate; D = drag-out rate; C_b = chemical concentration of the bath; C_r = chemical concentration in the rinse water; and t = time .

Q = 16 gal/min; D = 0.0099 gal; t = 3 min ; from these input data, the C_b/C_r ratio is solved for as:

C_b/C_r = Q (t/D) = 16 (3/0.0099) = 4848

This is approximately five times the industry norm. The plant can reduce its flow rate considerably and save money in the deionization of water and the treatment and disposal of its wastewater.

If the tanks are operated as flow through system with 96 min of rinse tank operation the water requirements will be:

96 min (16 gal/min) = 1,536 gal/d/rinse tank

If the system is operated in a batch rinse mode with 12 batches/d the water requirement will be:

30 gal/batch (12 batches/d) =360 gal/d/tank

8.4. *Case Studies Solution - 4 (continued).*

Batch operation represents over 75% reduction in water use.

Another approach to pollution prevention is to recycle the water. In addition, using flow controllers for water flow rates will minimize waste.

Evaluating cost: based on the assumption that the rinse water waste requires the same type of waste treatment with the chemicals as indicated above, the waste treatment cost will be reduced tremendously. With a total of seven rinse tanks, the amount of wastewater volume reduced/d will be:

(7 rinse tanks) (1,536 - 360 gal/d/tank) = 8,232 gal/d

The costs saved on chemicals are as follows:

Ferrous sulfate: (85 lb/1000 gal) (8232 gal/d)/(1000) $0.33/lb
= $230.9/d
Alum: (85 lb/1000 gal) (8232 gal/d)/(1000) $0.47/lb = $ 328.9/d
Sodium hydroxide: (20 lb/1000 gal) (8232 gal/d)/(1000) $0.74/lb
= $121.8/d
Polyelectrolyte: (0.3 lb/1000 gal) (8232 gal/d)/(1000) $7.50/lb
= $18.5/d
Total savings in chemicals = $700.10/d

Based on 350 days of operation/yr the plant can save $250,000/yr in chemicals alone!

Further, the amount of deionized water used for the rinse tanks will be reduced by 75% which will save additional dollars. Reduced waste treatment liability is an additional benefit.

8.5. *Case Studies Solution - 5.*

a. First, calculate the output of collectors based upon the data given.

Area of solar collectors (Ave. solar radiation (min)) (Efficiency)
= 1730 m^2 (154 watts/m^2 (December)) (0.91) (1 kW/1000 w)
= 242 kW

8.5. *Case Studies Solution - 5 (continued).*

The maximum cooling load demand is in September at 652 Btu/hr/d. The energy requirement for this month is:

652 Btu/hr/d (30 d/month) = 19,560 Btu/hr/month

Convert Btu to kW for comparison:

19,560 Btu/hr/month (1 watt/3.413 Btu/hr) = 5731 watts/month
= 5.7 kW/month

242 kW are available but only 5.7 kW are needed in a worst case scenario (December solar radiation in the peak load month, September.) The system would therefore only have to run at the following efficiency.

Efficiency = cooling power needed/cooling power available (100)
= 5.7 kW/242 kW (100) = 2.3 %

b. The first step is to calculate the cooling needs in cal/hr:

Cooling need (cal/hr) = Cooling load (Btu/hr) (watt/Btu/hr) (J/hr/watt) (cal/J)
= 652 Btu/hr (1 watt/3.413 Btu/hr) (3,600 J/hr/watt) (1.00 cal/4.184 J)
= 164,370 cal/hr = 164.4 kcal/hr

Next, the energy absorbed in raising the temperature of 1 gal of water by 10° F must be calculated, i.e., the energy required = heat capacity of water (g of water in 1 gal) (ΔT between input and output flow streams in C°). Note that the volume of the system is not relevant to this calculation.

Heat Absorption = (1.00 cal/g-°C) (1 gal) (3.78 L/gal) (1,000 g/L) (10 °F) (5/9 °C/°F)
= (1.00 cal/g-°C) (567,000 g) (5.55° C) = 20,970 cal/gal
= 21.0 kcal/gal

Next, the required flow rate is determined by dividing the cooling need by the heat absorption rate:

Pumping rate = (164.3 kcal/hr)/(21.0 kcal/gal) = 7.8 gal/hr

Finally, the pumping rate is convert to gal/min:

8.5. *Case Studies Solution - 5 (continued).*

Pumping rate = 7.8 gal/hr (1 hr/60 min) = 0.13 gal/min.

c. The total yearly cost for each system is:

LiBr system = $250,000 + $35,000 + $850 = $285,850/yr the first year; $285,850 + N ($850) each subsequent year, where N = number of years in operation after the first year

Conventional system = $200,000 + $20,000 + $375 + 36,500 kW/yr ($0.15/yr) = $220,375 + $5,475 = $225,850/yr for the first year; $225,850 + N ($5,475) each subsequent year, where N = number of years in operation after the first year

The payback period is calculated by setting each cost equal to one another and solving for N:

$$\$285,850 + N (850) = \$225,850/yr + N (\$5,475)$$

Solving for N, the payback period is found to be 13 years. This solar system cannot be justified on economic grounds, but may be found cost effective if environmental benefits are considered.

Some possible solutions may lie in designing the LiBr system to also provide space heating in the winter and to be used for hot water needs. It appears to have excess capacity and so perhaps a smaller unit would be more feasible (cheaper). In the case cited, NSF picked up a part of the bill and so it probably was a bargain.

d. The strategy for this problem is first to convert the temperature of 115 °F to °C and then to determine which of the four vapor pressure values for pure water is applicable. This value is then multiplied by the calculated value for the mole fraction of the water in the solution.

°C = (°F-32)(5/9) = (115 - 32) (5/9) = 46.0 °C

The relevant vapor pressure data from the table is 75.6 mm Hg (torr).

A 55% solution of LiBr contains 55 g of LiBr and 45 g of water in each 100 g. Molality is defined as the number of moles of solute per 1,000 g of solvent. The g of LiBr/1,000 g of water is calculated by use of a simple proportion:

8.5. *Case Studies Solution - 5 (continued)*.

$45/1,000 = 55/X$; $X = 1,222$ g of LiBr

1 gmol of LiBr has a mass of 86.85 g, therefore:

1,222 g LiBr (86.85 g/1 gmol of LiBr) = 14.1 gmol

The solution is 14.1 Molal.

The mole fraction is calculated as the number of moles of the substance in question (in this case, water) divided by the total number of moles in the system. For convenience, we can use the values calculated above:

Moles of water = 1000 g/(18.0 g/gmol) = 55.6 gmol
Moles of LiBr = 1,222 g/(86.85 g/gmol) = 14.1 gmol
The total number of gmol in the system = 55.6 + 14.1 = 69.7 gmol
The mole fraction for water in this system = 55.6/69.7 = 0.798

The vapor pressure over the system according to Raoult's Law is:

$P = 0.798$ (75.6 torr) =   60.3 torr = 60.3 mm Hg

In actuality, some deviation can be expected from this value since solutions are rarely ideal, especially at such high concentrations of solute.

8.6. *Case Studies Solution - 6*.

The station pumps 3,000 gal/d of gasoline, displacing 3,000 gal/d vapor. It operates 312 d/yr. The vapor mole fraction is 6.5/14.7 = 0.442. Ninety percent of the gasoline is recovered. A material balance can be used to calculate the mole fraction of the exhaust vapor for this 90% recovery rate. A material balance on the gasoline yields:

$$w_1 y_1 = w_2 y_2 + w_3 y_3$$

and a total material balance yields:

$$w_1 = w_2 + w_3$$

8.6. *Case Studies Solution - 6 (continued).*

where w_1 is the input moles, choose a basis of $w_1 = 1.0$, y_1 is the input mole fraction, 0.442, w_2 is the exhaust gas in moles, y_2 is the mole fraction of gasoline vapor in the exhaust gas, w_3 is the moles of gasoline recovered, 0.9 (w_1y_1), and y_3 is the mole fraction of the recovered gasoline, 1.0 (a pure liquid). Combining the two material balances to solve for y_2 yields:

$$y_2 = (w_1y_1 - w_3y_3)/w_2 = (w_1y_1 - 0.9\ w_1y_1)/(w_1 - 0.9\ w_1y_1)$$
$$= 0.1\ y_1/(1 - 0.9\ y_1) = 0.0734$$

and $w_2 = (w_1y_1 - w_3y_3)/y_2 = 0.1\ (0.442/0.0734) = 0.602$

The molar volume of the exhaust vapor is:

$$V/n = RT/P = 10.73\ (535)/14.7 = 390.6\ ft^3/lbmol$$

The total mole flow, w, is then:

$$w = 3{,}000\ gal/d\ (1\ ft^3/7.48\ gal)(1\ lbmol/390.6\ ft^3) = 1.027\ lbmol/d$$

The recovered gasoline is:

$$1.027\ lbmol/d\ (70\ lb/lbmol)\ (0.442)\ (0.9) = 28.6\ lb/d$$
$$= 5.53\ gal/d = 0.000662\ lb/s$$

The gas will need to be compressed to a pressure such that the gasoline is saturated at a mole fraction of 0.0734. Thus:

$$P_2 = 1\ (6.5)/(0.0734) = 88.6\ psia$$

The power required to compress to this pressure is:

$$P = 1.027\ lbmol/d\ (1\ d/30\ min)(1\ min/60\ s)\ (144\ in^2/ft^2)\ (10.73$$
$$lb_f/in^2\text{-}ft^3/lbmol\text{-}°R)\ (535\ °R)\ (\ln(88.6/14.7))/0.6$$
$$= 1412\ ft\text{-}lb_f/s = 1.91\ kW$$

The annual power cost is:

$$1.91kW\ (0.5\ hr/d)\ (312\ d/yr)\ (\$0.1/kW\text{-}hr) = \$29.80/yr$$

The annual income is:

$$5.53\ gal/d\ (312\ d\ /yr)\ (\$1.00/gal) = \$1{,}725.40/yr$$

The annual investment cost can be as much as:

8.6. *Case Studies Solution - 6 (continued).*

$1,725.4 − $29.80 = $1,695.60

The capital recovery factor is 0.1 $(1.1)^{10}/((1.1)^{10} -1) = 0.163$

Therefore the total capital investment could be:

$1,695.6/0.163 = $10,402

This is not enough money to justify the installation of much equipment. It would seem more feasible to let the tank truck put the vapor into the truck as the liquid is pumped out and then install a vapor recovery system at the bulk terminal where the volume will be larger.

8.7. *Case Studies Solution - 7.*

a. If G_{tot} is the total vapor emitted during the day, the equation given may be solved to obtain

$$C = G_{tot}/V$$

NOTE: It has been assumed that the concentration in the room is spatially homogeneous.

Since the vapor concentration is cumulative, it is appropriate to compute the mass of vapor which represents the short–term exposure limit value (the concentration to which a person should be exposed no longer than 15 minutes).

Using C = STEL = 200 ppm maximum concentration = 200×10^{-6} atm, and V = 80 m³, the total vapor mass allowed to be emitted daily is:

$$\frac{n}{V} = \frac{P}{RT} = \frac{(200 \times 10^{-6} \text{ atm})}{(82.057 \text{ cm}^3\text{-atm/gmol-K}) (298 \text{ K})} = 8.179 \times 10^{-9} \text{ gmol/cm}^3$$

$$= 8.179 \times 10^{-3} \text{ gmol/m}^3$$

$G_{tot} = 8.179 \times 10^{-3}$ gmol/m³ (131.39 g/gmol) (80 m³) = 85.97 g/8-h working day.

b. At steady state, C = G/kQ.

8.7. *Case Studies Solution - 7 (continued).*

Under these conditions, the TWA value should be used, since it represents the average concentration to which a worker can be continuously exposed over an 8–hour workday.

Using the ideal gas law, 50 ppm $= 0.269$ g/m^3.

Substituting into the steady-state solution,

$$0.269 \text{ g/m}^3 = 50 \text{ g/d} \left(\frac{1}{0.3 \text{ Q}}\right) (1 \text{ d/8 hr}) (1 \text{ hr/3600 s})$$

Rearranging to solve for Q yields:

$$Q = 0.022 \text{ m}^3\text{/s} = 1291 \text{ L/min}$$

as the minimum ventilation rate required to ensure that the TWA is not exceeded.

c. The total number of emissions sources is:

$$100 \text{ mi}^2 (2 \text{ dry cleaners/mi}^2) = 200 \text{ sources}$$

Each source emits 50 g/d, so the total is 10 kg/d = 3210 kg/yr = 6864 lb/yr operating 312 ·d/yr.

Compare this estimate with the estimate of 55.4 million pounds of TCE released in 1988 in the U.S. in 1988. Note that TCE is also widely used as a degreasing solvent in industrial applications.

d. First compute the vapor pressure at 25°C:

$$\log P = -\frac{1816.8}{298} + 7.95642 = 1.860; \quad P = 72.5 \text{ tor r}$$

Then use 50% of this value in the vapor pressure equation and solve for temperature:

$$\log (36.2) = -1816.8/(T_2) + 7.95642; \quad T_2 = 11°C$$

This temperature can reasonably be achieved. Similarly for 1% of this value:

$$\log (7.24) = -1816.8/(T_3) + 7.95642; \quad T_3 = -17.0°C$$

8.7. *Case Studies Solution - 7 (continued).*

e. Condensation occurs if the partial pressure in the cooled mixture equals the component vapor pressure at that temperature (assuming an ideal gas mixture):

50×10^{-6} atm (760 torr/atm) (0.25) = 9.5×10^{-3} torr
log (0.0095) = $-1816.8/T + 7.95642$
T = 182 K = $-90.9°C$

This is not a realistic temperature for an economical condensation unit. The vapors must first be concentrated, or compressed, or a different method of vapor recovery might be used. For example, it may be possible to cost–effectively separate the vapors using membrane separation techniques.

8.8. *Case Studies Solution - 8.*

The most efficient way of solving this problem is to use the method of Lagrange multipliers. This involves optimizing the total present value of the investment, subject to the constraint, $p \, l^4 = k$.

a. The total present value, TPV, equals the initial cost, p, plus the present value of the annual liability costs, l, which will be saved if the equipment is purchased, i.e.,

$$TPV = p + \beta \, l$$

where β is the present value function that depends upon the interest rate, the investment period, and the annual payment.

The Langrange multiplier method is formulated as follows:

$$L = p + \beta \, l - \lambda \, (k - p \, l^4)$$

Note that L is a function of p, l, and λ (the Lagrangian multiplier). To minimize L, the partial derivative of this expression is set equal to zero as follows:

$$\frac{\partial L}{\partial p} = 1 + \lambda \, l^4 = 0$$

8.8. *Case Studies Solution - 8 (continued).*

so that;

$$\lambda = -\frac{1}{l^4}$$

and,

$$\frac{\partial L}{\partial l} = \beta + 4 \lambda p l^3 = 0$$

so that;

$$p = \frac{\beta l}{4}$$

and,

$$\frac{\partial L}{\partial \lambda} = k - p l^4 = 0$$

so that;

$$l = \left(\frac{4 k}{\beta}\right)^{1/5}$$

The equation for p is now expressed in terms of known variables as follows:

$$p = \left(\frac{\beta}{4}\right) l = \left(\frac{\beta}{4}\right) \left(\frac{4 k}{\beta}\right)^{1/5} = \left(\frac{\beta}{4}\right)^{4/5} k^{1/5}$$

Since a numerical value for k is not given in the problem statement, values of $p/(k^{1/5}) = (\beta/4)^{4/5}$ and $l/(k^{1/5}) = (4/\beta)^{1/5}$ have been tabulated below.

Table 1. Values of $p/(k^{1/5})$ and $l/(k^{1/5})$ for a 20-year investment period and an interest rate of 5%.

Interest Rate	β	$p/(k^{1/5})$	$l/(k^{1/5})$
5%	12.462	2.48	0.80

8.8. *Case Studies Solution - 8 (continued).*

b. The calculations proceed similarly for the other interest rates, and are summarized in Table 2 for plotting in Figure 1. Note that the higher the interest rate, the lower the optimal equipment investment is. In addition, for higher equipment investments, the liability costs per year are reduced as expected from the functional relationship given in the problem statement.

Table 2. Values of $p/(k^{1/5})$ and $l/(k^{1/5})$ for a 20-year investment period and an interest rate of 5 to 25%.

Interest Rate	β	$p/(k^{1/5})$	$l/(k^{1/5})$
5%	12.462	2.48	0.80
10%	8.514	1.83	0.86
15%	6.259	1.43	0.91
20%	4.870	1.17	0.96
25%	3.954	0.99	1.00

b. The calculations proceed similarly for the 5-year investment period, and are summarized in Table 3 for plotting in Figure 1. Note that with a shorter investment period, the impact of the variable interest rate is not as apparent. Also, the optimal equipment investment is significantly smaller than that for the 20-year investment period case, with the result that the liability reduction is not as large (i.e., the liability costs per year are higher than in Table 2).

The inverse relationship between equipment purchased and liability costs is seen clearly in Figure 1.

Table 3. Values of $p/(k^{1/5})$ and $l/(k^{1/5})$ for a 5-year investment period and an interest rate of 5 to 25%.

Interest Rate	β	$p/(k^{1/5})$	$l/(k^{1/5})$
5%	4.329	1.07	0.98
10%	3.791	0.96	1.01
15%	3.352	0.87	1.04
20%	2.991	0.79	1.06
25%	2.689	0.73	1.08

8.8. *Case Studies Solution - 8 (continued).*

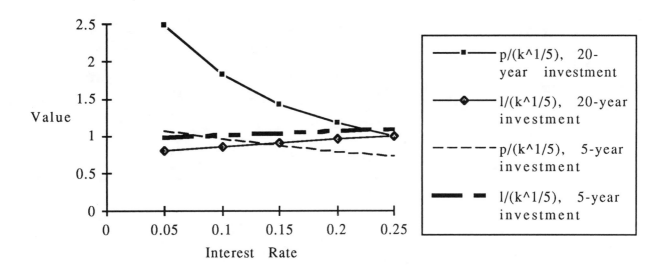

Figure 1.   Plot of optimal equipment investment and resultant liabilty costs for various investment periods and prevailing interest rates.

9. Ethics

9.1. *Ethics Solution - 1.*

Term	Explanation
Ethical theory	An attempt to answer certain questions about standards of conduct.
Consequentialist Theory	(Also referred to as teleological theory). This is an ethical theory to evaluate acts and policies where the rightness of an act is determined by its end.
Deontological Theory	This is a nonconsequentialist ethical theory. Any ethical theory which presents morality as a system of absolute duties is a deontological system. An example is the Ten Commandments.
Utilitarianism	This is the most influential consequentialist theory. The view is that an act is right in proportion to its increase of the sum of human happiness or decrease of unhappiness for all affected. A risk/benefit analysis, or a cost/benefit analysis are examples.
Ethical Egoism	This is a consequentialist ethical theory implying that the good is that which benefits the individual.
Nationalism	This is a consequentialist ethical theory implying that the good is that which benefits the state.
Retributivism	This is a deontological position that states that the criminal deserves to suffer solely because he or she has broken the law. The argument is that by breaking the law, the criminal acted unjustly, benefitted at other's expense, and failed to respect his or her fellow citizens. Therefore the criminal ought to be punished. An example is the principle of "a life for a life."

9.2. *Ethics Solution - 2.*

Traditionally, engineers and scientists search for the best solution to technical problems, e.g., designing a system to produce the highest quality medicine, producing gasoline at the lowest price, minimizing expenses, maximizing profits, etc.

Existentialism is the search of the inner truth, or the elimination of external influence. Engineers usually have existential freedom in searching for the best solution to a technical problem. This is acceptable, until the question of social and environmental responsibility is raised. It may be economically advantageous to have a degreasing unit open to the air, if pollution prevention is not a factor. However, waste minimization and health risks necessitate the use of VOC recovery systems.

It is often stated that engineers care. But the question is for what?

As with ethical theories, the attempt to answer questions about ethics does not directly provide answers to specific questions.

9.3. *Ethics Solution - 3.*

Failure to report a spill could be a violation of a law, rule, or regulation. It could result in fines or even criminal prosecution. The focus of this question, however, is strictly on the ETHICAL aspects of the company's actions.

The following are some of the ethical considerations of this action. You may have thought of others.

While the company's prompt action in correcting the problem is praiseworthy, its failure to report it is unethical. Environmental regulatory programs are largely self-policing. They rely heavily on voluntary reporting. Failure to report such a problem lessens the integrity of the entire system and gives a false picture of the need to protect the environment.

Failure to report the release also places an unfair burden on other companies that use the same chemical. That is because the sewer authority, having observed high concentrations of the chemical, places additional monitoring and treatment requirements on all the

9.3. *Ethics Solution - 3 (continued).*

companies. Had the company in question reported the releases, the other companies would not have had to implement these additional, expensive, monitoring and treatment requirements.

Sewer workers and members of the public may have been harmed by direct exposure to the toxic chemicals. Exposure could continue for some time even after the releases have ended. If the chemical is odorless and tasteless, the releases may have gone undetected. Failure to report the releases may deprive these people of the ability to protect themselves from further exposure and from receiving proper medical help.

Even though the releases have stopped, much damage may have occurred to sewage treatment plants, streams, rivers, fish and other aquatic life. People may have been harmed by eating contaminated fish, clams, lobsters and the like. It is unethical not to inform the public of such dangers. If a cleanup is required to remediate any damage caused to the environment, it would be unethical to remain silent while the taxpayers foot the bill.

9.4. *Ethics Solution - 4.*

This is an open-ended question with many possible answers. The following are some of the arguments that may be made.

BENEFITS

The employee is highly motivated (as a means of self protection) to learn the law and to be very careful not to violate it. Managers are forced to scrutinize the activities of the people they supervise. It often results in all actions being thoroughly documented in order to avoid any liability. The policy also helps employees resist pressures to cut corners whether that pressure comes from managers, or from the need to get a job done more quickly and/or less expensively.

DRAWBACKS

That the government will not defend its employees implies that the employees are guilty of any violations with which they may be charged even before they are convicted. Many employees find such a situation to be demoralizing. Employee loyalty may be reduced. Some employees may become nervous and overly cautious - seeking

9.4. *Ethics Solution - 4 (continued).*

> to have every detail of what they do approved in writing. Day-to-day business gets tied up in red tape as written legal opinions are sought at every step.

> Violations, even those made as honest errors, may, out of fear of prosecution, be hidden and never reported. Such a punitive program could be a powerful disincentive to critical self-auditing, self-policing, and voluntary disclosure.

SUGGESTED IMPROVEMENTS
Many ways to improve the program are possible. A few of them are:

1. Limit the policy of placing the employee's actions outside the scope of his or her employment to incidents of:
   a. deliberate violations,
   b. violations that occurred after having been warned, and
   c. gross negligence.
   This would greatly reduce the disincentive to critical self-auditing, self-policing, and voluntary disclosure.
2. Balance the punitive aspect of the incentive program with positive aspects such as giving awards for reporting environmental problems and/or correcting them.
3. Reduce the need for red tape documentation of the legality of every action with frequent and intensive training programs in environmental compliance.
4. Allay the employee's fears of environmental inspections by having joint meetings of the employees with the inspectors and management at regular intervals.

2.9.5. *Ethics Solution - 5.*

> This is an open-ended question with many possible answers. Encourage vigorous discussion from all points of view.

Some important points, however, must always be kept in mind:

- It is never ethical to violate the law or to advise someone to violate the law.

9.5. *Ethics Solution - 5 (continued).*

- It cannot be ethical for the farmer to do nothing as others may be harmed by the spilled fuel.

- Financial loss to the farmer may be tragic, but that cannot make it ethical to take actions that could cause harm to others.

9.6. *Ethics Solution - 6.*

This is an another open-ended question with many possible answers. Encourage vigorous discussion from all points of view, but encourage critical thinking.

9.7. *Ethics Solution - 7.*

This is an open-ended question with many possible answers. Encourage vigorous discussion from all points of view. Students should consider the costs of environmental compliance, the cost in human life and increased health care burdens, the ethics of U.S. companies having two sets of standards - one within our borders and one outside our borders, etc.

9.8. *Ethics Solution - 8.*

This is an open-ended question with many possible correct answers.

Dr. Smith's theory might be used to argue against pollution prevention laws by claiming that dilution of pollutants by the atmosphere or oceans down to the parts per million range is sufficient protection of people and the environment. In fact, such exposure is probably actually good for you while the expense of implementing regulations aimed at reducing the concentrations of those chemicals even further would be so expensive as to put companies out of business and people out of work. The unemployed workers and their families would suffer far more from poor nutrition and lack of health care than from the exposure to chemicals at a concentration that proved to be beneficial to the animals in the test. It could also be argued that those expensive (and according to Dr. Smith's theory,

9.8. *Ethics Solution - 8 (continued).*

unnecessary) regulations may deprive large numbers of people of the benefits of the products made with the polluting chemicals because the regulations may make it either impractical to produce or expensive beyond their ability to pay.

Factors that would favor continuance of pollution abatement laws and programs could include the following:

1. The very small dose of toxic chemical that actually improves health would vary greatly from person to person. In a large population such as in the United States, even a very low percentage of adversely affected people translates into a large number of people harmed by the chemical. Thus, there would be incentive to keep concentrations of that toxic chemical in the environment as low as possible.
2. Experimentally determining the range of doses of toxic chemical that produces the beneficial "sufficient challenge" would be extremely difficult and prohibitively expensive.
3. Toxic chemical testing rarely determines the toxic effect of mixtures of chemicals. Since the population would always be exposed to mixtures of the tested chemical with other chemicals in the environment, it would not be possible to confidently say what dose would actually be beneficial. Again, the incentive would be to keep the dose of each toxic chemical, in the mixture known as pollution, as low as possible.
4. There is no guarantee that any humans will have a beneficial "sufficient challenge" based on a test animal's reaction. The dose, therefore, should be kept as low as possible.

9.9. *Ethics Solution - 9.*

State and federal governments are showing an increasing propensity towards criminal prosecution of violators of environmental regulations. The courts have generally upheld the concept of a "corporate veil" that protects everyone except those with direct knowledge of environmental violations. In most cases where corporate executives have been successfully prosecuted, they have been directly involved in the day-to-day management of hazardous materials. Recently, a new tactic has been employed whereby the

9.9. *Ethics Solution - 9 (continued)*.

> top executives are notified of violations by corporate counsel, preferably outside counsel. These communications are shielded by the attorney-client privilege and, as such, are not subject to review by the courts. As Director of Environmental Affairs, you should be concerned about this type of arrangement since it may build an effective legal barrier leaving you on the outside. A corporate structure that prevents direct reporting of violations to management should be questioned.

9.10. *Ethics Solution - 10*.

> The ethical dilemma in this problem arises from the decision whether or not to save money by utilizing a recycling operation that the government has attempted to characterize as a "sham". The difference between the charge for "recycling" ($1,282/ton) and the market price of the product ($35/ton) raises questions as to whether the process is recycling or incineration. A more complete investigation into the operation of the kiln and the EPA charges could resolve any doubts regarding this issue.

> The argument made in favor of utilizing the "recycler" option is that it may possess less potential environmental risk than land disposal and no greater risk than conventional incineration (at a lower cost). If the process truly reduces the long-term risk of human exposure, then the environmental manager would be forced to choose the best method regardless of the "sham" characterization (a rose by any other name still smells as sweet).

> Persuasive arguments can be made on both sides and these are left to classroom discussion.